CLIMATE CHANGE
The Point of No Return

There can be no doubt: climate change is in full swing and man's influence on the climate is increasing. This book offers a solid scientific portrayal of the state of the climate debate. Rather than horror scenarios or trivialization, this work describes and assesses the issues as objectively as possible. Latif addresses the basics of climate physics, describes the state of scientific knowledge regarding the causes of global climate change, and discusses the climate changes that are already observable today. He counters the conventional "skeptics' arguments" and outlines how our future climate could look if humanity does not subscribe to sustainability. The work's conclusion presents concrete options for action.

Mojib Latif, professor of meteorology and climate physics at the Leibniz Institute of Marine Sciences at the University of Kiel, is one of the best-known meteorologists and climate experts in Germany. He has published numerous studies on climate change. He has been awarded the Max Planck Award for Public Science, for his ability to explain complicated research results in an understandable way.

This volume has been updated to be consistent with the new IPCC report, published in 2007.

Our addresses on the Internet:
www.the-sustainability-project.com
www.forum-fuer-verantwortung.de
[English version available]

CLIMATE CHANGE
The Point of No Return

MOJIB LATIF

Translated by Susan E. Richter

Klaus Wiegandt, General Editor

HAUS PUBLISHING

First published in Great Britain in 2009 by
Haus Publishing Ltd
70 Cadogan Place
London SW1X 9AH
www.hauspublishing.com

Originally published as: *Bringen Wir das Klima aus dem Takt?* by
Mojib Latif

A CIP catalogue record for this book
is available from the British Library

ISBN 978-1-906598-14-3

Typeset in Sabon by MacGuru Ltd
Printed in Dubai by Oriental Press

CONDITIONS OF SALE

Mixed Sources
Product group from well-managed
forests and other controlled sources
www.fsc.org Cert no. CU-COC-809367
© 1996 Forest Stewardship Council

FSC

Haus Publishing believes in the importance of a
sustainable future for our planet. This book is
printed on paper produced in accordance with the
standards of sustainability set out and monitored by
the FSC. The printer holds chain of custody.

Contents

Editor's Foreword

Sustainability Project

Sales of the German-language edition of this series have exceeded all expectations. The positive media response has been encouraging, too. Both of these positive responses demonstrate that the series addresses the right topics in a language that is easily understood by the general reader. The combination of thematic breadth and scientifically astute, yet generally accessible writing, is particularly important as I believe it to be a vital prerequisite for smoothing the way to a sustainable society by turning knowledge into action. After all, I am not a scientist myself; my background is in business.

A few months ago, shortly after the first volumes had been published, we received suggestions from neighboring countries in Europe recommending that an English-language edition would reach a far larger readership. Books dealing with global challenges, they said, require global action brought about by informed debate amongst as large an audience as possible. When delegates from India, China, and Pakistan voiced similar concerns at an international conference my mind was made up. Dedicated individuals such as Lester R. Brown and Jonathan Porritt deserve credit for bringing the concept of sustainability to the attention of the general public, I am convinced that this series can give the discourse about sustainability something new.

Two years have passed since I wrote the foreword to the initial German edition. During this time, unsustainable developments on our planet have come to our attention in ever more dramatic ways. The price of oil has nearly tripled; the value of industrial metals has risen exponentially and, quite unexpectedly, the costs of staple foods such as corn, rice, and wheat have reached all-time highs. Around the globe, people are increasingly concerned that the pressure caused by these drastic price increases will lead to serious destabilization in China, India, Indonesia, Vietnam, and Malaysia, the world's key developing regions.

The frequency and intensity of natural disasters brought on by global warming have continued to increase. Many regions of our Earth are experiencing prolonged droughts, with subsequent shortages of drinking water and the destruction of entire harvests. In other parts of the world, typhoons and hurricanes are causing massive flooding and inflicting immeasurable suffering.

The turbulence in the world's financial markets, triggered by the US sub-prime mortgage crisis, has only added to these woes. It has affected every country and made clear just how unscrupulous and sometimes irresponsible speculation has become in today's financial world. The expectation of exorbitant short-term rates of return on capital investments led to complex and obscure financial engineering. Coupled with a reckless willingness to take risks everyone involved seemingly lost track of the situation. How else can blue chip companies incur multi-billion dollar losses? If central banks had not come to the rescue with dramatic steps to back up their currencies, the world's economy would have collapsed. It was only in these circumstances that the use of public monies could be justified. It is therefore imperative to prevent a repeat of speculation with short-term capital on such a gigantic scale.

Taken together, these developments have at least significantly improved the readiness for a debate on sustainability. Many more are now aware that our wasteful use of natural resources and energy have serious consequences, and not only for future generations.

Two years ago, who would have dared to hope that WalMart, the world's largest retailer, would initiate a dialog about sustainability with its customers and promise to put the results into practice? Who would have considered it possible that CNN would start a series "Going Green?" Every day, more and more businesses worldwide announce that they are putting the topic of sustainability at the core of their strategic considerations. Let us use this momentum to try and make sure that these positive developments are not a flash in the pan, but a solid part of our necessary discourse within civic society.

However, we cannot achieve sustainable development through a multitude of individual adjustments. We are facing the challenge of critical fundamental questioning of our lifestyle and consumption and patterns of production. We must grapple with the complexity of the entire earth system in a forward-looking and precautionary manner, and not focus solely on topics such as energy and climate change.

The authors of these twelve books examine the consequences of our destructive interference in the Earth ecosystem from different perspectives. They point out that we still have plenty of opportunities to shape a sustainable future. If we want to achieve this, however, it is imperative that we use the information we have as a basis for systematic action, guided by the principles of sustainable development. If the step from knowledge to action is not only to be taken, but also to succeed, we need to offer comprehensive education to all, with the foundation in early childhood. The central issues of the future must be anchored firmly in school curricula, and no university student should be permitted to graduate without having completed a general course on sustainable development. Everyday opportunities for action must be made clear to us all – young and old. Only then can we begin to think critically about our lifestyles and make positive changes in the direction of sustainability. We need to show the business community the way to sustainable development via a responsible attitude to consumption, and become active within our sphere of influence as opinion leaders.

For this reason, my foundation *Forum für Verantwortung*, the ASKO EUROPA-FOUNDATION, and the European Academy Otzenhausen have joined forces to produce educational materials on the future of the Earth to accompany the twelve books developed at the renowned Wuppertal Institute for Climate, Environment and Energy. We are setting up an extensive program of seminars, and the initial results are very promising. The success of our initiative "Encouraging Sustainability," which has now been awarded the status of an official project of the UN Decade "Education for Sustainable Development," confirms the public's great interest in, and demand for, well-founded information.

I would like to thank the authors for their additional effort to update all their information and put the contents of their original volumes in a more global context. My special thanks goes to the translators, who submitted themselves to a strict timetable, and to Annette Maas for coordinating the Sustainability Project. I am grateful for the expert editorial advice of Amy Irvine and the Haus Publishing editorial team for not losing track of the "3600-page-work."

Taking Action — Out of Insight and Responsibility

"We were on our way to becoming gods, supreme beings who could create a second world, using the natural world only as building blocks for our new creation."

This warning by the psychoanalyst and social philosopher Erich Fromm is to be found in *To Have or to Be?* (1976). It aptly expresses the dilemma in which we find ourselves as a result of our scientific-technical orientation.

The original intention of submitting to nature in order to make use of it ("knowledge is power") evolved into subjugating nature in order to exploit it. We have left the earlier successful path with its many advances and are now on the wrong track, a path of danger with incalculable risks. The greatest danger stems from the unshakable faith of the overwhelming majority of politicians and business leaders in unlimited economic growth which, together with limitless technological innovation, is supposed to provide solutions to all the challenges of the present and the future.

For decades now, scientists have been warning of this collision course with nature. As early as 1983, the United Nations founded the World Commission on Environment and Development which published the Brundtland Report in 1987. Under the title *Our Common Future*, it presented a concept that could save mankind from catastrophe and help to find the way back to a responsible way of life, the concept of long-term environmentally sustainable use of resources. "Sustainability," as used in the Brundtland Report, means "development that meets the needs of the present without compromising the ability of future generations to meet their own needs."

Despite many efforts, this guiding principle for ecologically, economically, and socially sustainable action has unfortunately not yet become the reality it can, indeed must, become. I believe the reason for this is that civil societies have not yet been sufficiently informed and mobilized.

Forum für Verantwortung

Against this background, and in the light of ever more warnings and scientific results, I decided to take on a societal responsibility with my foundation. I would like to contribute to the expansion of public discourse about sustainable development which is absolutely essential. It is my desire to provide a large number of people with facts and contextual knowledge on the subject of sustainability, and to show alternative options for future action.

After all, the principle of "sustainable development" alone is insufficient to change current patterns of living and economic practices. It does provide some orientation, but it has to be negotiated in concrete terms within society and then implemented in patterns of behavior. A democratic society seriously seeking to reorient itself towards future viability must rely on critical, creative individuals capable of both discussion and action. For this reason, life-long learning, from childhood to old age, is a necessary precondition for realizing sustainable development. The practical implementation of the ecological, economic, and social goals of a sustainability strategy in economic policy requires people able to reflect, innovate and recognize potentials for structural change and learn to use them in the best interests of society.

It is not enough for individuals to be merely "concerned." On the contrary, it is necessary to understand the scientific background and interconnections in order to have access to them and be able to develop them in discussions that lead in the right direction. Only in this way can the ability to make appropriate judgments emerge, and this is a prerequisite for responsible action.

The essential condition for this is presentation of both the facts and the theories within whose framework possible courses of action are visible in a manner that is both appropriate to the subject matter and comprehensible. Then, people will be able to use them to guide their personal behavior.

In order to move towards this goal, I asked renowned scientists to present in a generally understandable way the state of research and the possible options on twelve important topics in the area of sustainable development in the series *"Forum für Verantwortung."* All those involved in this project are in agreement that there is no alternative to a united path of all societies towards sustainability:

- *Our Planet: How Much More Can Earth Take?* (Jill Jäger)
- *Energy: The World's Race for Resources in the 21st Century* (Hermann-Joseph Wagner)
- *Our Threatened Oceans* (Stefan Rahmstorf and Katherine Richardson)
- *Water Resources: Efficient, Sustainable and Equitable Use* (Wolfram Mauser)
- *The Earth: Natural Resources and Human Intervention* (Friedrich Schmidt-Bleek)
- *Overcrowded World? Global Population and International Migration* (Rainer Münz and Albert F. Reiterer)
- *Feeding the Planet: Environmental Protection through Sustainable Agriculture* (Klaus Hahlbrock)
- *Costing the Earth? Perspectives on Sustainable Development* (Bernd Meyer)
- *The New Plagues: Pandemics and Poverty in a Globalized World* (Stefan Kaufmann)

- *Climate Change: The Point of No Return* (Mojib Latif)
- *The Demise of Diversity: Loss and Extinction* (Josef H Reichholf)
- *Building a New World Order: Sustainable Policies for the Future* (Harald Müller)

The public debate

What gives me the courage to carry out this project and the optimism that I will reach civil societies in this way, and possibly provide an impetus for change?

For one thing, I have observed that, because of the number and severity of natural disasters in recent years, people have become more sensitive concerning questions of how we treat the Earth. For another, there are scarcely any books on the market that cover in language comprehensible to civil society the broad spectrum of comprehensive sustainable development in an integrated manner.

When I began to structure my ideas and the prerequisites for a public discourse on sustainability in 2004, I could not foresee that by the time the first books of the series were published, the general public would have come to perceive at least climate change and energy as topics of great concern. I believe this occurred especially as a result of the following events:

First, the United States witnessed the devastation of New Orleans in August 2005 by Hurricane Katrina, and the anarchy following in the wake of this disaster.

Second, in 2006, Al Gore began his information campaign on climate change and wastage of energy, culminating in his film *An Inconvenient Truth*, which has made an impression on a wide audience of all age groups around the world.

Third, the 700-page Stern Report, commissioned by the British government, published in 2007 by the former Chief Economist of the World Bank Nicholas Stern in collaboration with other economists, was a wake-up call for politicians and business leaders alike. This report makes clear how extensive the damage to the global economy will be if we continue with "business as usual" and do not take vigorous steps to halt climate change. At the same time, the report demonstrates that we could finance countermeasures for just one-tenth of the cost of the probable damage, and could limit average global warming to 2° C – if we only took action.

Fourth, the most recent IPCC report, published in early 2007, was met by especially intense media interest, and therefore also received considerable public attention. It laid bare as never before how serious the situation is, and called for drastic action against climate change.

Last, but not least, the exceptional commitment of a number of billionaires such as Bill Gates, Warren Buffett, George Soros, and Richard Branson as well as Bill Clinton's work to "save the world" is impressing people around the globe and deserves mention here.

An important task for the authors of our twelve-volume series was to provide appropriate steps towards sustainable development in their particular subject area. In this context, we must always be aware that successful transition to this type of economic, ecological, and social development on our planet cannot succeed immediately, but will require many decades. Today, there are still no sure formulae for the most successful long-term path. A large number of scientists and even more innovative entrepreneurs and managers will have to use their creativity and dynamism to solve the great challenges. Nonetheless, even today, we can discern the first clear goals we must reach in order to avert a looming catastrophe. And billions of consumers around the world can use their daily purchasing decisions to help both ease and significantly accelerate the economy's transition to sustainable development – provided the political framework is there. In addition, from a global perspective, billions of citizens have the opportunity to mark out the political "guide rails" in a democratic way via their parliaments.

The most important insight currently shared by the scientific, political, and economic communities is that our resource-intensive Western model of prosperity (enjoyed today by one billion people) cannot be extended to another five billion or, by 2050, at least eight billion people. That would go far beyond the biophysical capacity of the planet. This realization is not in dispute. At issue, however, are the consequences we need to draw from it.

If we want to avoid serious conflicts between nations, the industrialized countries must reduce their consumption of resources by more than the developing and threshold countries increase theirs. In the future, all countries must achieve the same level of consumption. Only then will we be able to create the necessary ecological room for maneuver in order to ensure an appropriate level of prosperity for developing and threshold countries.

To avoid a dramatic loss of prosperity in the West during this long-term process of adaptation, the transition from high to low resource use, that is, to an ecological market economy, must be set in motion quickly.

On the other hand, the threshold and developing countries must commit themselves to getting their population growth under control within the foreseeable future. The twenty-year Programme of Action adopted by the United Nations International Conference on Population and Development in Cairo in 1994 must be implemented with stronger support from the industrialized nations.

If humankind does not succeed in drastically improving resource and energy efficiency and reducing population growth in a sustainable manner – we should remind ourselves of the United Nations forecast that population growth will come to a halt only at the end of this century, with a world population of eleven to twelve billion – then we run the real risk of developing eco-dictatorships. In the words of Ernst Ulrich von Weizsäcker: "States will be sorely tempted to ration limited resources, to micromanage economic activity, and in the interest of the environment to specify from above what citizens may or may not do. 'Quality-of-life' experts might define in an authoritarian way what kind of needs people are permitted to satisfy." (*Earth Politics*, 1989, in English translation: 1994).

It is time

It is time for us to take stock in a fundamental and critical way. We, the public, must decide what kind of future we want. Progress and quality of life is not dependent on year-by-year growth in per capita income alone, nor do we need inexorably growing amounts of goods to satisfy our needs. The short-term goals of our economy, such as maximizing profits and accumulating capital, are major obstacles to sustainable development. We should go back to a more decentralized economy and reduce world trade and the waste of energy associated with it in a targeted fashion. If resources and energy were to cost their "true" prices, the global process of rationalization and labor displacement will be reversed, because cost pressure will be shifted to the areas of materials and energy.

The path to sustainability requires enormous technological innovations. But not everything that is technologically possible has to be put into practice. We should not strive to place all areas of our lives under the dictates of the economic system. Making justice and fairness a reality for everyone is not only a moral and ethical imperative, but is also the most important means of securing world peace in the long term. For this reason, it is essential to place the political relationship between states and peoples on a new basis, a basis with which everyone can identify, not only the most powerful. Without common principles of global governance, sustainability cannot become a reality in any of the fields discussed in this series.

And finally, we must ask whether we humans have the right to reproduce to such an extent that we may reach a population of eleven to twelve billion by the end of this century, laying claim to every square centimeter of our Earth and restricting and destroying the habitats and way of life of all other species to an ever greater degree.

Our future is not predetermined. We ourselves shape it by our actions. We can continue as before, but if we do so, we will put ourselves in the biophysical straitjacket of nature, with possibly disastrous political implications, by the middle of this century. But we also have the opportunity to create a fairer and more viable future for ourselves and for future generations. This requires the commitment of everyone on our planet.

Klaus Wiegandt
Summer 2008

Author's Foreword

The influence humans exert on the climate is increasing. The global climate change we have initiated is a consequence of the wide range of human activities that release trace gases into the atmosphere. These gases include carbon dioxide, methane and nitrous oxide, all of which have effects on the climate. The energy sector, agriculture, and industrial production stand out as the most prominent emitters. An additional factor is deforestation. The amplification of the so-called 'greenhouse gases' in the atmosphere since the beginning of industrialization is leading to a global warming that brings with it far-reaching changes in the Earth system. This is the essence of the climate problem. Yet the link between the content of trace gases in the atmosphere and the global climate has been known for over 100 years.

The climate problem is not an isolated issue; it must be regarded in the context of the other major problems facing mankind. The climate problem is inseparable from demographic growth, for instance, and intimately connected with the energy question. The availability of water is crucially dependent on climate, as is biodiversity. However, the complex of climate problems also has obvious socio-economic aspects, which extend far into society, into the private sector, and into politics. This is why I jumped at the chance to participate as an author in The Sustainability Project series, which deals with the twelve central themes of the future, interwoven with each other in a great variety of ways. The overarching idea is the principle of sustainability. Only if the policies of national states and the world commit to the idea of sustainability will the enormous challenges facing humanity be resolved in our globalized world. In the long term, the increased use of renewable energies is especially important in addressing the climate problem, as the continued use of fossil energy resources will lead to dramatic climate changes in coming years. Because of the large inertia of the climate, complete restructuring into a carbon-free global economy can take place gradually over this century without causing any major economic dislocations. Therefore it is up to us to stabilize our climate at a halfway acceptable level. One of this author's concerns is to communicate this possibility to civil society.

I would like to thank the *Forum für Verantwortung*, especially Mr. Klaus Wiegandt, who proposed this book project and lent it generous support. Further thanks are due to my fellow authors for the quite inspiring discussions at the authors' meetings, as well as to my colleagues at the Leibniz Institute of Marine Sciences in Kiel and at the Max Planck Institute for Meteorology in Hamburg. Last but not least, I thank my wife Elisabeth for her endless patience while I worked on this book in my free time.

Kiel, 7 July 2008

The Scientific Principles

1 The Climate and the Earth System

1.1 Weather and Climate

The difference between weather and climate can be explained in one sentence, uttered by Larry Gates, one of the American pioneers of climate research, at an event held by the World Meteorological Organization (WMO) in the early 1980s: 'Climate is what you expect, weather is what you get.' So there is a fundamental difference between weather and climate. Weather research is concerned with the generation, propagation and prediction of individual weather elements, such as a certain mid-latitudinal storm or a hurricane, while climate research is interested in the entirety of mid-latitudinal storms and hurricanes, and addresses such questions as how many storms or hurricanes there will be next year, or whether they will become more frequent or intensify as a consequence of global warming. With the term 'weather' we thus designate the short-term occurrences in the atmosphere, while the term 'climate' refers to longer periods of time. The World Meteorological Organization defines climate as the statistics of weather over a period that is long enough to allow these statistical properties to be determined. While weather describes the physical state of the atmosphere at a certain point of time at a certain location, a description of climate is not complete until it includes the probability for deviations from the mean, and thus the probability of extreme weather events occurring, such as the flooding of the Elbe river in 2002 or the drought in Germany the following year. Generally a span of thirty years is used as the reference period for a description of the climate.

The term 'climate' is derived from *klinein*, the Greek word for 'tilt,' since summer and winter are consequences of the inclination of the Earth's axis relative to the plane containing the Earth's orbit around the sun, known as the ecliptic. Currently this tilt measures 23.5°, meaning that the Northern Hemisphere is irradiated more strongly during the northern summer and the Southern Hemisphere more strongly during the northern winter. For the most part, the usual geographical classification into climatic zones follows the resulting seasonal cycle of meteorological values such as temperature and precipitation. The tilt of the Earth's axis, along with other so-called orbital parameters, vary over the course of millennia and are in part responsible for the fact that the history of the Earth has seen strong climatic excursions(see Section 3.5).

The different seasonal and annual mean levels of the sun's irradiation at the equator and the poles results in great regional differences in the surface temperature of the Earth. The resulting horizontal temperature differences in the lower atmosphere lead to differences in atmospheric pressure. Finally, the resulting force of pressure, together with the influence of gravity and the rotation of the Earth, causes the winds. The average three-dimensional wind pattern is called the 'general circulation' of the atmosphere. The atmosphere is not an isolated system, however, but interacts with other components of the earth system: with the hydrosphere (oceans and the water cycle on continents and in the atmosphere), the cryosphere (ice and snow), the biosphere (animals and plants), the pedosphere (soil) and the lithosphere (rock). These components define the climate system (see Figure 1), and they all move at completely different

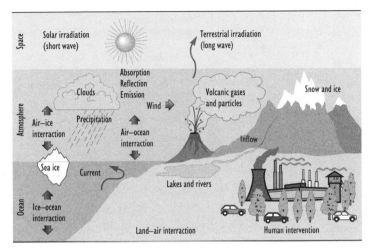

Figure 1   Schematic diagram of the Earth's climate system

speeds and exhibit drastically different heat conductivities and heat capacities.

The dynamics of the climate system and the climate statistics that follow from it are thus characterized by the extremely different time scales of its individual components (see Figure 2). While the lower atmosphere adapts to conditions at the surface in just hours, the deep ocean circulation takes many centuries to adjust to changed boundary conditions like a modified composition of the atmosphere, and a large ice sheet like the Antarctic ice sheet requires many millennia for the process. Changes in the climate can emerge from interactions within a single climate component, or from interactions among the individual components, for instance between the ocean and the atmosphere. Changes can also be triggered externally by such events as a change in solar irradiation, by volcanic eruptions, or by a change in the composition of the Earth's atmosphere. In the past hundred years man

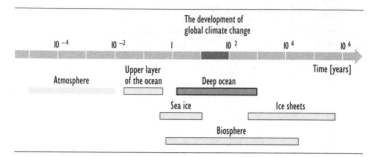

Figure 2   The time scales characteristic for the individual components
of the climate system

has taken on ever greater importance for the climate, by emitting into the atmosphere trace gases affecting the climate, which changes the atmospheric radiation budget, and thus contributes to global warming on the Earth.

1.2 The Oceans

Figure 2 shows a diagram of the components active in the climate system. It illustrates the corresponding time scales, ranging from hours up to millions of years. The 'fastest' component is the atmosphere, in which the significant energetic processes (such as the sequence of high and low pressure systems in our latitudes) exhibit a typical time scale of several days. The water cycle, which makes a crucial contribution to atmospheric energetics by means of the heat flows arising from evaporation and precipitation, is estimated to have a period of about 10 days. On the opposite end of the scale in Figure 2 are the ice sheets of Greenland and the Antarctic, whose characteristic time scales are in the range of many millennia. Fluctuations of this kind correspond to the rhythms of the global warm and cold periods (e.g. ice ages), which occur at intervals of between 10 000 and many 100 000 years.

Between the two extremes in Figure 2 lies the ocean, divided into the layers close to the surface and the deep sea. The former react over periods of weeks and months, as the currents and stratification fluctuate due to direct links between the surface and the varying atmospheric fields of wind, temperature, radiation and precipitation. Changes in the deep sea, by contrast, depend on fluctuations in the surface conditions in limited regions of the polar and subpolar latitudes. Because of the huge amounts of water involved, they play out over periods of time ranging from decades to several centuries. Since global warming refers to precisely this range of the time scales, the oceans play a very important role in anthropogenically induced climate change.

In continuous media, different time scales are also associated with different spatial scales. The spatial scale significant for atmospheric energetics occurs at around 1000 km. This corresponds to the typical extension of the high and low pressure systems in the atmosphere. In the ocean the most energetic current and stratification changes extend to distances of around 10 to 100 km. These changes take the form of 'mesoscale' vortexes (eddies), the oceanic analog to atmospheric highs and lows. The 'fast' atmospheric processes (which take several days) have a comparatively larger extension (1000 km) than the 'slow' processes (which can last for months) within the mixed layer of the ocean (the 'mixed layer' is the top layer, well mixed by waves, tides and weather events). This makes the observation of the most energetic oceanic movements much more difficult than that of the atmospheric ones. Observations show that mesoscale eddies exist even at great depths of several thousand meters. Because these, too, play an important role for long-term climate fluctuations, they must be subjected to constant observation and simulated in models.

The central role of water in the climate system is based on the asymmetrical structure of the water molecule, which makes it an electric dipole, encouraging the temperature-dependent aggregation of water molecules. This is why water has the anomalous property of possessing its greatest density at 4°C, such that its solid phase, ice, floats. We find huge amounts of floating ice, known as 'sea ice,' in the Arctic and Antarctic. Because its molecules cohere, water reacts inertly to both warming and cooling, lending it the highest heat capacity of all liquid and solid substances (with the exception of ammonia), and shifting its boiling and freezing points to 100°C and 0°, respectively, from the −80°C and -110°C that would be expected if the water molecule had a symmetrical structure.

Moreover, water is an extremely effective solvent. The relative weight component of the dissolved substances is called 'salinity,' and averages 3.47% in the world's oceans (in oceanography salinity is usually given in tenths of a percent). This salinity significantly changes the above mentioned properties of water. For instance, at a salinity of 3.47% the temperature of the maximum density shifts to −3.8°C, and drops below the freezing point of −1.9°C. This means that the phenomenon of 'convection' can take place while the ocean cools until ice forms: cooled water sinks, warmer water rises from the depths, releasing its heat content to the atmosphere, and then sinks to the depths after absorbing atmospheric gases. Convection plays a major role in cloud formation in the atmosphere as well. However, it also takes place in the interior of the Earth and is thus one of the most important processes propelling movement throughout the entire Earth system.

With regard to circulation and stratification (the vertical

structure of density), primary stimulation of the ocean can take place only from the surface. In the case of a wind forcing, the interplay of friction, the deflective force of the Earth's rotation (Coriolis force) and the shape of the ocean basin lead to the system of surface currents known from any atlas, whose intensity decreases with depth. Familiar examples include the Gulf Stream in the Atlantic and the Kuroshio in the Pacific. Since the wind fields are neither temporally nor spatially homogenous, the mass transport associated with the currents causes regional mass surpluses or deficits. The rise and drop in the ocean surface associated with these effects drive not only long period waves and horizontal currents in the interior of the ocean, but also vertical motions. Upwelling regions are frequently nutrient-rich areas and thus tend to feature strong biological production.

As the ocean surface warms or cools, or as salinity changes due to evaporation, precipitation or the formation of sea ice, water masses of different density are created. In the Tropics and Subtropics, the surface layers are warmed constantly in the annual mean. The correspondingly light surface water floats as a warm top layer – with a temperature above 10°C – on the water masses located below. The strong temperature difference (density difference) quite effectively prevents the warming from penetrating the deeper layers, so that the top layer remains relatively thin, with a thickness between 100 and 800 meters. In the subpolar and polar areas the average heat deficit results in an increase in density, reinforced by the salt released when seawater freezes in certain regions; on average, the effect is deep convection.

Correspondingly, the oceans of higher latitudes are characterized by low stratification; signals from the surface can reach the interior of the ocean quite effectively and vice versa. The cold water from the polar oceans – with a temperature of less than 10°C – penetrates under the warm top layer into the interior:

This launches a circulation driven by temperature and salt (the thermohaline circulation), for the cold water flowing toward the equator as it sinks to the depths is compensated for by the poleward transport of warm water in the top layer. The resulting oceanic heat transport on the Northern Hemisphere measures a maximum of $3.0 * 10^{15}$ Watts; this amount, combined with the heat transport of the atmosphere, is needed to balance the Northern Hemisphere's radiation budget of $5.5 * 10^{15}$ Watts. Applied to the Atlantic, this thermohaline circulation is generally referred in media to as the 'Gulf Stream circulation' or simply the 'Gulf Stream.'

The thermohaline circulation is propelled by convection processes in the higher latitudes. The surface conditions define the characteristics of the cold water masses created. These characteristics can be followed along their path toward the deep sea by measuring their temperatures, salinities, and densities or the content of oxygen and other trace gases. Most of the deepest water masses are of Antarctic origin (known as Antarctic Bottom Water). The water masses of Arctic origin are slightly less dense and thus stratified over the bottom layer. Since only the Atlantic Ocean extends to the higher latitudes on the Northern Hemisphere, this water begins its circulation in the Greenland Sea and the Labrador Sea (North Atlantic Deep Water). Figure 3 shows a diagram of its global spreading and the compensatory return current in the top layer. This circulation, also known as the 'global conveyor belt' or the 'global overturning circulation,' has a cycle lasting several hundred up to a thousand years, as estimated using dating methods measuring the number of radioactive carbon isotopes in deep water (isotopes are different types of atoms of the same chemical element each having different atomic mass).

The longest part of this cycle takes place in the cold path of the conveyor belt in the deep layers, where typical velocities

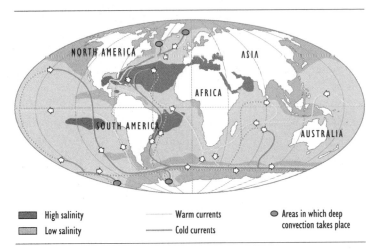

High salinity   Warm currents   Areas in which deep
Low salinity   Cold currents   convection takes place

Figure 3   Schematic diagram of the thermohaline circulation

amount to just 1 to 3 km per day. The great masses of water involved in this circulation, amounting to around 0.4 billion km³ (around one third of the entire amount of seawater) transported at a rate of around 20 million m³/s, combined with this water's heat storage capacity and high solubility, makes the thermohaline circulation one of the most important long-term warehouses in the climate system. The substances stored in this warehouse include some of the greenhouse gas carbon dioxide emitted by man. Because the biogeochemical processes involved are still largely unknown, it is not possible to determine the capacity of the oceanic warehouse accurately. Therefore it is uncertain how long the ocean will continue to curb the atmospheric rise of anthropogenic CO_2 and thus moderate the global warming caused by an increased greenhouse effect (see section 2.4).

A crucial question pertaining to regional and global climate change concerns the stability of the thermohaline circulation. This issue trains on the changeability of convection in the higher latitudes. On the one hand, convection could be reduced by warming as a consequence of the anthropogenic greenhouse effect. On the other hand, in low-temperature areas salinity variations are very important, like those caused by changes in continental ice sheets, in precipitation and in river runoff from the continents into the North Atlantic – processes that also can be influenced by anthropogenic climate change. This complex of issues, known in the media as the 'Gulf Stream problem,' is a matter of intensive discussion not only in science, but in the general public as well. Even Hollywood took on this issue in the movie *The Day After Tomorrow* (see section 4.5). Suffice it to say at this juncture that a collapse of the thermohaline circulation will not trigger a new ice age.

1.3 Sea Ice

Sea ice is part of the cryosphere. This sphere encompasses all forms of snow and ice, making it the second largest component of the climate system in terms of mass and heat capacity. The significant components of the cryosphere are snow, sea ice, mountain glaciers, the ice shelves and the continental ice sheets. Today ice permanently covers 10% of the land's surface (14.8 million km^2) and around 6.5% of the oceans in the annual mean (22.5 million km^2). With an area of 1.5 million km^2, the ice shelves, most of which are in Antarctica, cover a significantly smaller ocean surface than does sea ice. However, the volume of the ice shelf is around 0.66 million km^3, over ten times greater than that of sea ice. All mountain glaciers and smaller ice caps together have a volume of around 0.18 million km^3.

Variations in the extent, solidity, thickness and drift of sea ice arise through dynamic and thermodynamic processes. The thermodynamics of sea ice, i.e. freezing and melting, is influenced by radiation processes and by the turbulent fluxes of latent and sensible heat. Sea ice drift is caused by wind stress, ocean current, Coriolis force and internal stresses in sea ice (deformation). The internal stresses are determined using rheological models. Rheology is a discipline of solid state physics and continuum mechanics that deals with the flow and deformation of materials.

Changes in the sea ice boundary are among the most important properties of climate fluctuations in the Polar Regions. On the geophysical scale, sea ice is a thin, broken layer on the polar oceans, which is moved by wind and ocean currents and whose thickness and extent is affected by thermodynamic processes. Sea ice constitutes the boundary between the two much larger components of the Earth system, atmosphere and ocean, and thus exerts considerable influence on their interaction. In March sea ice covers 5% of the ocean surface; in September 8%. In the Arctic Ocean the average thickness is 3 m, in the Antarctic, 1 m. Since sea ice, even if it is not covered by snow, has quite a high albedo (capacity to reflect solar radiation) and thus a significant influence on the radiation budget, it plays an important role in the climate system. This role is further increased by the fact that its insulating effect prevents any direct, turbulent exchange of momentum and heat between the ocean and the atmosphere. Thus the atmosphere is considerably colder above the surface of sea ice than over the open sea. The occurrence of sea ice not only cools the air in the Polar Regions, but also increases the meridional temperature gradient at the Earth's surface – the temperature contrast between the Tropics and the Polar Regions – and thus intensifies the westerly winds in the mid-latitudes.

Sea ice influences the ocean as well as the atmosphere. The

insulation of the ocean from heat losses to the cold atmosphere, and the change in stress at the ocean surface through losses in momentum due to the deformation of sea ice, are two important processes conditioned both by the intensive reduction in turbulent exchange and by the behavior of sea ice as a viscous-plastic solid flowing in two dimensions, driven by wind and ocean currents. Another important effect is the influence of sea ice on convection in the ocean and thus on the formation of deep and bottom water. Sea ice therefore also plays an important role in large-scale ocean circulation, especially for the thermohaline circulation. The average salinity of sea water is 3.47%, but that of sea ice is only around 0.5‰. When sea water freezes, a considerable quantity of salt is released into the ocean. This raises the density of sea water, making it heavier, and can trigger convection. The process of sea ice formation supports the generation of dense sea water in the Polar Regions, which can sink to deeper ocean layers thereby affecting the ocean circulation.

Some climate fluctuations on the time scale of decades are explained in the literature by means of the concept of 'ice-ocean oscillators.' For instance, imagine a fluctuation that includes the thermohaline circulation, in which increased ice formation leads to higher salinity and thus density of the water in the North Atlantic. The result is more convection at higher latitudes, causing the thermohaline circulation to intensify. Thus more heat is transported northward in the North Atlantic, causing sea ice to melt. As a consequence, more freshwater enters the ocean, reducing the density at its surface, with the result that convection – and thus the thermohaline circulation – is weakened. The effect is a decreased northward heat transport, ultimately causing an increase in ice formation. Such interactions between ocean and sea ice are important in understanding natural climate variability, especially on the time scale of decades.

Yet sea ice also plays an important role in connection with the anthropogenic greenhouse effect. All climate models that simulate the effect of increased concentrations of greenhouse gases yield the result of accelerated warming in the Arctic. Indeed, the last decades have seen a reduction in Arctic sea ice, in terms of both average thickness and spatial extent. In this the phenomenon of ice albedo feedback plays a decisive role. Light-colored surfaces like ice have an extremely high albedo, which means that they reflect a considerable portion of incident sunlight back into space; accordingly, less solar energy remains on Earth as a source of heat. When sea ice retreats as a consequence of the anthropogenic greenhouse effect, this means that less sunlight is reflected and more energy is available, leading to additional warming. This positive feedback is one of the reasons why the Arctic is warming up particularly rapidly. The retreat of Arctic sea ice in the 20th century can be simulated using climate models that account for the observed concentrations of greenhouse gases. The models also show that a further rapid reduction of Arctic sea ice is expected in the next decades. Should concentrations of greenhouse gases continue to climb so strongly, the Arctic may even be completely free of ice in summer by the end of the century. The sea ice around the Antarctic reacts less sensitively to human influence, as one consequence of the stronger vertical mixing processes in the Southern Ocean is reduced surface warming in this region. Melting sea ice does not contribute to the rise in sea levels, as it floats and displaces a corresponding volume of water even in its solid form.

1.4 The Continental Ice Sheets

The continental ice sheets and ice shelves constitute the much larger part of the cryosphere by volume. The term 'continental ice sheet,' also simply called 'ice sheet' refers to an ice mass of continental scale resting on land, which formed over the course of millennia through the accumulation of snow. Ice shelves are ice masses which are both connected to an ice sheet and floating in the ocean. Ice shelves are fed primarily by the flow of an ice sheet's mass toward the ocean. They typically exist in the large bays of a continental shelf that are covered by ice sheets. By far the largest body of ice on our Earth is the Antarctic ice sheet, with a total volume of 24.7 million km^3 of ice, an ice-covered surface area of 12.3 million km^2 and a mean thickness of 2.1 km (not including the ice shelves). If the Antarctic ice sheet were to melt completely, sea levels would rise by a good 60 m. The attached ice shelves have a volume of 0.66 million km^3, an area of 1.5 million km^2 and a mean thickness of 0.44 km. If these melt it would not change the sea level, as it, like sea ice, is already displacing water by floating in the ocean.

The Greenland ice sheet is the second largest ice sheet on Earth. Its volume amounts to around 3 million km^3 and it is around 1.7 million km^2 in area. Complete melting of the Greenland ice sheet would correspond to a sea level rise of about 7 m globally. Since there are no large bays, the Greenland ice sheet calves off directly into the ocean; no significant ice shelves exist. An important difference from the Antarctic ice sheet is that the areas near the edge and located at greater depths melt because of higher surface temperatures in the summer. Therefore the Greenland ice sheet is much more vulnerable to the multiple-degree rises in temperature expected as a consequence of the anthropogenic greenhouse effect than is the Antarctic ice sheet.

In the past the Earth has seen periods with considerably more glaciation. At the pinnacle of the last ice age 21, 000 years ago, the areas covered by ice sheets included major portions of North America and northern Europe. At this time the entire volume of ice was about three times greater than today, and, accordingly, the global sea level was about 130 m lower. The fact that Great Britain and continental Europe were connected by dry land is a good illustration of what this meant. Two factors will play a particularly important role in the future. For one, in the long term the Milankovitch theory predicts a new ice age, which will reach its pinnacle in 50 000 years at the earliest. Second, in the short term, over a period of 100 to 1000 years, we expect a clear reduction in glaciation as a result of the anthropogenic greenhouse effect.

Continental ice sheets represent an important dynamic component of the Earth system over very long time scales ranging from several centuries to many millennia. Ice sheets possess distinctive viscous flow properties. They are therefore not merely rigid, immobile solids that change their form through snowfall and melting, but actually diverge under their own weight in order to maintain a dynamic equilibrium. The loss of mass in the interior is counteracted by the accumulated snowfall, while the outflow at the edge is compensated for by melting and calving. Ice sheets thus have a complex interrelationship with the adjacent systems of atmosphere, ocean and lithosphere. The mass balance of ice sheets is controlled by atmospheric parameters like precipitation, air temperature and radiation, which determine snowfall and melting behavior. Feedback mechanisms influencing the atmosphere emerge as a consequence of changes in their topography (shape) and albedo.

The ice sheet and the ocean, too, are subject to reciprocal effects. The surrounding ocean affects an ice sheet in the sense that the sea level determines the part of a land mass accessible to glacier formation and the position of the ice shelf. The melting or calving of ice constitutes a source of freshwater to the ocean, through which ice sheets feed back on the large-scale

ocean circulation. Melted sea water can refreeze on the under-
side of an ice shelf; as most of the salt remains in the water,
this water is extremely saline and dense, causing it to sink to
great depths. Geothermal heat flows into the ice sheet, where it
rests on the rock. Because of this thermal input, the temperature
of an ice column generally increases with depth, such that near
the ground it may reach the melting point. This, in turn, produces more water, further reducing ice viscosity. Under certain
circumstances a lubricating film may result, along which the
ice sheet may slide down into the ocean. A mechanism of this
kind is considered a real possibility for the West Antarctic ice
sheet, where such a rapid destabilization would result in global
sea levels rising by 5 to 6 m. The weight of the ice sheet lowers
the rock below it as well, by levels of up to around 30% of the
thickness of the ice sheet. Having been relieved of the burden of
the glaciers of the last ice age, North America and Scandinavia
continue to rise even today.

Because they are so massive, ice sheets need an extremely long
time to respond to changes in climate. A typical time scale can be
calculated from the ratio of the characteristic ice thickness and
the characteristic accumulation rate of snow. Assuming an ice
thickness of 2 km and an accumulation rate of 100 mm/year in
the Antarctic, this time scale amounts to 20 000 years; for Greenland, which has a characteristic ice thickness of 1.5 km and an
accumulation rate of 300 mm/year, 5000 years. It follows from
this that the ice sheets can be regarded more or less as a constant
over the next hundred years. Beyond the time horizon of several
centuries, however, the dynamics of the ice sheets must be taken
into consideration. Suffice it to say here that how long it will take
before the ice sheets are destabilized is a matter of controversial
discussion among scientists. From paleoclimatic reconstructions,
we know today that a great deal of ice entered the oceans within

a relatively short time at the end of the last ice age, which made sea levels rise by up to 10 m within just a few centuries.

Current calculations using global climate models predict possible scenarios of a temperature rise of 6 to 10°C over the ice sheets within this century, which could continue over many centuries. Simulations with complex ice sheet models, assuming a temperature rise of 10°C, show the volume of the Greenland ice sheet shrinking by at least one third after 1000 years, which would correspond to a good 2 m rise in sea level. Because of significantly lower surface temperatures, the Antarctic ice sheet turns out to be considerably more stable. A corresponding simulation for the Antarctic returned decreases in volume of a good 5%; however, due to the great absolute volume of ice, this corresponds to a sea level rise of around 3.5 m. It must be emphasized, however, that these simulations are rough estimations at best, since several processes are not taken into consideration. For instance, the assumption is that the soot dumped into the atmosphere by us humans does not influence the albedo of the ice sheets. Yet soot could reduce the albedo of the ice sheets considerably, and, in so doing, further reinforce the melting of the ice sheets in summer and their infiltration with liquid water, thus accelerating their erosion, with the effect that the rise in sea levels would accelerate significantly.

1.5 Vegetation

A further important component in the Earth system is vegetation. Vegetation interacts with the atmosphere's momentum budget; it influences the Earth's hydrological cycle and the earth's radiation budget, and has an important influence on the cycles of materials like carbon. Vegetation is not static, but closely interrelated to the other components of the Earth system, and has been a major contributor to several climate changes in the past. The future climate, too, will be influenced decisively by the behavior of vegetation. The interactions between vegetation and the chemical composition of the Earth's atmosphere are referred to as 'biogeochemical feedback mechanisms.' The interactions between vegetation and the energy, water and momentum budgets are referred to as 'biogeophysical feedback mechanisms'.

Probably the most important biogeochemical feedback mechanism concerns the interaction between vegetation and the concentration of carbon dioxide in the atmosphere. This feedback can be either negative or positive, and is dependent in part on the time scale of the process observed. Imagine that biomass increases for some non-specified reason. This event results in more CO_2 being removed from the atmosphere. Through this, the greenhouse effect is reduced further and the air temperature near the ground drops. This effect slows plant growth, damping the initial disturbance, and is thus a negative feedback. Yet the anthropogenic greenhouse effect could also cause positive feedback. As a consequence of major changes in the climate, entire forest regions could experience so much strain that they take up less CO_2 than they return to the atmosphere through respiration. This could further amplify the anthropogenic greenhouse effect. Several simulations with models show that this effect, averaged over all land regions, could contribute additional warming of around 1°C by the end of the 21st century. Recently an additional biogeochemical feedback connected with the methane cycle was discovered. Plants apparently produce methane, a feedback that was not previously known. The impact of this process on vegetation's climate effect overall is a focus of current research. However, it is to be expected that the calculations on the anthropogenic greenhouse effect will not be influenced significantly by this newly discovered mechanism.

One of the most important biogeophysical interactions concerns the change in albedo and thus the absorption of solar irradiation as a consequence of changes in vegetation. This feedback is positive. Since vegetation generally has a lower albedo than non-vegetated ground, as vegetation spreads the absorption of solar irradiation will increase and thus raise ground-level temperatures. This favors further plant growth. The feedback via albedo is especially strong in areas with great albedo contrasts, that is, in desert regions or in regions periodically covered with snow. The bright sand desert of the Sahara, for instance, has an albedo of around 0.5, while the bordering savannah typically has an albedo of around 0.2. Even greater contrasts are found in the high northern latitudes. Over the snowy forests of the taiga, typical values are around 0.3, while much higher values of up to 0.9 are measured in grassy areas covered with snow.

In addition to the albedo, plants also influence the water budget of the atmosphere at ground level. Most of this effect is due to transpiration from leaf stomata and uptake of water through roots. These hydrological interactions can exhibit both positive and negative feedback mechanisms. The interplay between the various biophysical feedback mechanisms is quite complex. However, various experiments suggest that, averaged over the year, the albedo effect tends to dominate in the high northern latitudes, while the hydrological effect prevails in the Tropics. Considering the biogeochemical and the biophysical processes together – as is essential in the study of global climate change – is generally possible only with relatively complete Earth system models. Simulations of this kind show that biogeochemical feedback dominates in the Tropics, while in the intermediate and higher latitudes of the Northern Hemisphere biogeophysical feedback prevails. Therefore deforestation in the Tropics presumably results in global warming, and reforestation in global cooling. The conditions are exactly reversed in the intermediate and higher latitudes of the Northern Hemisphere, so that there deforestation results in cooling and reforestation in warming. However, complexes of such kinds of questions only recently have found consideration in models of the Earth system, so that not even the direction of certain feedback mechanisms is known with any certainty.

Many climate model simulations further suggest that the feedbacks from vegetation dynamics can amplify or accelerate externally stimulated climate changes. Paleoclimatic studies using climate models show that highly non-linear interactions of the physical climate system (atmosphere, ocean, sea ice) with vegetation have made a decisive contribution to the changes observed in the past. One example of this is the phenomenon of the 'green Sahara.' The extremely fast transition from a Sahara with lush vegetation to a desert around 6000 years ago can only be reproduced with models when biogeophysical feedback mechanisms involving vegetation are incorporated. The physical cause, namely the gradual change in solar irradiation during summer in the Northern Hemisphere as a consequence of the variation in Earth's orbital parameters, is not sufficient to explain the speed of the shift. Moreover, models of the physical climate system alone cannot explain the north-south extent of vegetation before the Sahara became a desert. Although there are still many gaps in our understanding of the role of vegetation for the climate, it is clear that vegetation is not just a passive climate indicator, but an interactive component in the Earth system, without which the dynamics of the climate cannot be fully explained.

2 The Greenhouse Effect

2.1 The Composition of the Atmosphere

The optimum living conditions on Earth are due in part to the chemical composition of Earth's atmosphere, which is clearly different from that of other planets in the solar system. The main components of Earth's atmosphere are nitrogen (N_2) at 78%, and oxygen (O_2) at 21%, which thus make up around 99% (Figure 4). But we also owe our life-nurturing climate on Earth to a few other gases that occur only in minute traces – hence the name trace gases – but nevertheless exert a strong influence on the Earth's climate by influencing the planet's radiation budget. The most important of these include water vapor (H_2O), carbon dioxide (CO_2) and ozone (O_3). Carbon dioxide, for instance, currently has a share of only 0.0384% (384 ppm, ppm = parts per million), but is of major importance for our present and future climate.

Measurements prove beyond a doubt that the manifold activities of humans are changing the composition of the atmosphere, in particular by raising the concentration of the persistent trace gases relevant for the climate (Figure 5, Table 1). This is the core of the climate problem. The concentrations of carbon dioxide (CO_2), methane (CH_4) and nitrous oxide (N_2O) are rising tremendously over their preindustrial levels. There are a great many reasons for this, including the strong rise in the consumption of

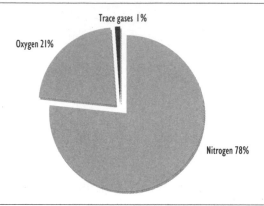

Figure 4 The composition of the atmosphere
Particularly important for the climate are the trace gases, which make up only
one percent of the atmosphere

fossil energy sources (oil, natural gas, and coal), the expansion
of industrial production, changed land usage, and the expansion of animal husbandry. The demographic explosion also plays
an important role. Even substances that do not occur at all in
nature have entered the atmosphere, such as chlorofluorocarbons
(CFC), which are created only by humans. Many of the trace
gases humans have emitted into the atmosphere are so persistent
that they have the potential to influence the climate for centuries. Combined with the inertia of the climate, which reacts quite
slowly to disturbances, this makes for a long-term problem that
will occupy many successive generations.

Figure 5 The development of the three trace gases carbon dioxide (CO_2),
methane (CH_4) and nitrous oxide (N_2O) over the last 1000 years
The corresponding radiative forcing is also given. The rise in concentrations since
the beginning of industrialization can be recognized clearly (from IPCC 2001a)

Global concentration of three well-mixed greenhouse gases in the atmosphere

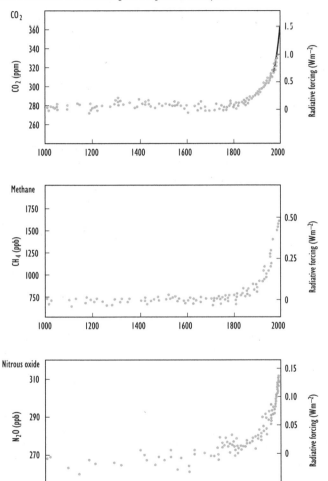

These trace gases, which also include water vapor and ozone, exhibit a special property. They let the radiation from the sun (above all in the visible, shortwave range) that falls on the Earth pass through the atmosphere nearly unchecked, but absorb part of the radiation emitted back to space by the warmed Earth (in the non-visible, long-wave range). This is the essence of the 'greenhouse effect.' The gases involved are thus generally called 'greenhouse gases.'

	CO_2 carbon dioxide	CH_4 methane	N_2O nitrous oxide	CFC-11 freon-11	HFC-23 fluoroform	CF4 terafluor-methane
Preindustrial concentration	280 ppm	700 ppb	270 ppb	0	0	0
Concentration 2005	379 ppm	1774 ppb	319 ppb	251 ppt	18 ppt	74 ppt
Change since 1998	+13 ppm	+11 ppb	+5 ppb	−13 ppt	+4 ppt	0
Atmospheric lifetime in years	~100 (5–200)	12	114	45	270	50000

Table 1   The most important anthropogenic trace gases and several of their characteristics (according to IPCC 2001a)
The units ppm, ppb and ppt mean parts per million, parts per billion and parts per trillion

As mentioned above, the greenhouse gases emitted into the atmosphere by man have relatively long lifetimes. For carbon dioxide the lifetime fluctuates widely depending on the process of removal, but averages around one hundred years. What is more, the rates of change in concentrations also fluctuate within certain boundaries. The rate of CO_2 increase, for instance, accelerated from 1.5 ppm/year in the 1990s to about 2 ppm/year in the period 1998–2007. Because of their long lifetimes, trace gases scatter across the planet, making their effects felt globally, regardless of where they were emitted. Since the existing gas sinks are not sufficient to completely rid the atmosphere of the greenhouse gases released by humans, their concentrations rise. This process is akin to the accumulation of public debt, which climbs ever higher through the continuous addition of new debt: Even reducing new debt by a few percent would not keep the burden from growing. A minor reduction in the emission of greenhouse gases would have a correspondingly minor effect: the concentrations of most of the trace gases would still continue to rise, and thus their effect on the climate would grow as well.

2.2 The Natural Greenhouse Effect

Discussions of the climate problem often involve the catchword 'greenhouse effect' – a natural property of the Earth's atmosphere. This phenomenon is the guarantee for our optimum living conditions on Earth. As is known, matter emits electromagnetic radiation of all wavelengths in the form of photons (emissions), and these emissions increase with the heat of the emitting matter. So, for instance, the emission of electromagnetic radiation on the (hot) surface of the sun provides for energy in the form of visible light on the Earth (between around 0.2 and 5 μm). However, incidental electromagnetic radiation is also swallowed by matter (absorption), so that additional energy enters the environment, generally felt in the form of warming. In the temperatures of the terrestrial climate system, which are obviously much lower than those on the sun, most of the electromagnetic radiation emitted by the Earth's surface and/or atmospheric components takes place in the non-visible, 'thermal' range of the spectrum (between around 3 and 100 μm), and thus is often referred to as 'thermal radiation.'

It is also important for the greenhouse effect that emission and absorption can be highly dependent on the wavelength, especially for gases, which is why the terms 'emission lines' or 'bands' (aggregations of lines) are often used. Absorption and emission by the two important greenhouse gases water vapor (H_2O) and carbon dioxide (CO_2), in particular, take place in such bands. In contrast, the two main gases in the atmosphere, oxygen (O_2) and nitrogen (N_2) exhibit no significant emission or absorption in the energetic range of the spectrum. Thus the trace gases are the main culprits that exert a major influence on the terrestrial climate.

On an Earth without atmosphere, the surface temperature would be determined exclusively by the balance between irradiated solar energy and the thermal radiation sent out by the planet's surface. At today's backscatter ratio (albedo), the global average surface temperature would average around −18°C, but instead this temperature is currently around +15°C (see Figure 6). Even an atmosphere of pure oxygen and nitrogen – after all, the main components of our atmosphere – would not make any significant change: Our planet would be an icy desert, and life as we know it probably never would have emerged.

Water vapor, however, and to a somewhat lesser extent, CO_2 (and other trace gases) also absorb a small amount of solar radiation and emit thermal radiation themselves. In the downward direction this additional thermal radiation from the atmosphere is greater than the solar radiation absorbed, making the energy input at the surface higher than it would be without such gases. This process further warms the surface of the Earth and (as a consequence of various transport processes) the lower atmosphere as well.

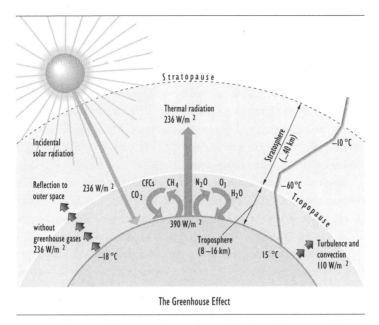

The Greenhouse Effect

Figure 6   Schematic diagram of the radiation budget of the atmosphere
and the natural greenhouse effect

The warming of the Earth's surface also results in an adjust-
ment to the radiation balance at the upper edge of the atmosphere, however, for averaged over the longer term the thermal
radiation the Earth emits into space must be equivalent to the
radiation it absorbs from the sun. The energy radiated upward
from the Earth's surface is also absorbed (in part) by the atmospheric trace gases, so that only part of it ends up directly in
space. These gases emit energy themselves as a function of their
temperature, yet since their temperatures fall with increasing
elevation in the atmosphere, this energy is much lower than at
the Earth's surface. Therefore if the temperature at the Earth's

surface remains constant, and the quantity of trace gases increases, ever less energy in the form of thermal radiation leaves the Earth's surface toward outer space. However, as surface temperatures rise, this deficit in the radiation budget is balanced by the Earth's surface emitting correspondingly more thermal radiation. An important auxiliary in this is the atmospheric radiation window, a spectral range at a wavelength of about 10 μm within which radiation from the surface can escape through a cloudless atmosphere into space.

Satellite measurements of thermal radiation in space yield the conclusion that the temperature of the Earth's surface increases by about 33°C due to the natural greenhouse effect. Water vapor (H_2O) makes by far the greatest contribution to this life-sustaining warming – approximately two thirds; next follow carbon dioxide (CO_2) with a share of around 15%, ozone with approximately 10%, and finally nitrous oxide (N_2O) and methane (CH_4) with about 3% each. Therefore, despite their low concentrations, the existence of the trace gases is one of the decisive factors determining our climate. One often hears that the human contribution to the greenhouse effect amounts to only around 2%, from which some draw the conclusion that the influence of man on the climate is negligible. But if the greenhouse effect contributes over 30°C, 2% amounts to around 0.6°C. This more or less corresponds to the share of warming over the past one hundred years that is attributed to humans in climate simulation models. This so-called 'skeptic's argument' thus confirms the results of climate research. Number games of this kind thus must always be challenged critically and calculated through to the end.

The overall effect of trace gases is thus to trap heat in the lower atmosphere. In a manner of speaking, the atmosphere is not transparent for thermal radiation, but for the most part permeable for solar radiation. Because of the analogy with the processes in a greenhouse, where the glass also lets the sunlight through but not the heat, the phenomenon described here is also known as the 'natural greenhouse effect.' The gases in the atmosphere responsible for this effect are thus frequently called greenhouse gases. Basically, these greenhouse gases play the role of the glass in a greenhouse. In interpreting various climate processes, the greenhouse image must not be applied all too directly, however. The physical processes in a real greenhouse are entirely different from those that take place in the atmosphere. Moreover, the conditions in a moving atmosphere with clouds are, of course, much more complicated than those in a gardener's glass house.

2.3 The Anthropogenic Greenhouse Effect

When the naturally occurring greenhouse gases (e.g., CO_2) are increased through anthropogenic (human) influence, or new greenhouse gases like the chlorofluorocarbons (CFC) are added, this has to have an impact on our climate. An increased concentration of greenhouse gases in the atmosphere, as a consequence of the enhanced (anthropogenic) greenhouse effect, necessarily leads to an increase in the temperature of the Earth's surface and the lower atmosphere. Other factors that also can have an influence on the climate, like, for instance, the condensation trails ('contrails') caused by airplanes and the emission of aerosols, will not be considered in detail at this juncture (see section 4.3). Figure 7 shows the 'radiative forcings' of the various external (anthropogenic and natural) factors that have been influencing our climate since the beginning of industrialization. Radiative forcings are generated by changes in the composition of the

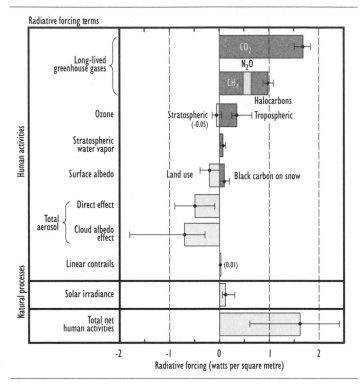

Figure 7   Summary of the principal components of the radiative forcing

The values represent the forcings in 2005 relative to the start of the industrial era (about 1750). Human activities cause significant changes in long-lived gases, ozone, water vapor, surface albedo, aerosols and contrails. The only increase in natural forcing of any significance between 1750 and 2005 occurred in solar irradiance. Positive forcing leads to a warming of the climate and negative forcing to a cooling. The thin black line attached to each colored bar represents the range of uncertainty for the respective value (redrawn after IPCC 2007)

atmosphere, changes in surface albedo through land usage, and irradiation fluctuations by the sun. Radiative forcings caused in the short term by volcanoes are not represented.

The concentration of the persistent greenhouse gases, as mentioned above, is increasing systematically: since the beginning of industrialization up to today, carbon dioxide has increased by approximately 30%, methane by approximately 150% and nitrous oxide by approximately 17%. The main case of this trend, responsible for around half of this rise, is the use of fossil fuels. Around 20% of the entire emission of greenhouse gases worldwide comes from chemical production. The most important of these are the CFCs. Another important source of globally emitted greenhouse gases is increasingly intensive agricultural production, responsible for 15% of emissions. The destruction of forests makes up another 15%. The radiative forcing resulting from the rise in greenhouse gases caused by man since the preindustrial era is positive, amounting to around 2.8 W/m², which has to lead to warming. The radiative forcing of the aerosols is negative, and thus leads to cooling. However, the latter is smaller than the radiative forcing from anthropogenic greenhouse gases. The sum of all factors including the solar effect is positive and amounts to 1.6 W/m², and in this the effect of anthropogenic greenhouse gases clearly dominates. During the second half of the 20th century the positive radiative forcing of the well mixed greenhouse gases rose rapidly; in contrast, the sum of all natural radiative forcings was negative. Thus humans are well on their way to triggering a massive climate disruption in the form of global warming. The first sign of this is the rapid warming of the Earth over the last thirty years (see Figure 18), of a magnitude never experienced in the past two millennia.

The positive radiative forcing initiates a long-term warming of the Earth's surface and the lower atmosphere, the extent of which increases in keeping with the change in the atmosphere's greenhouse gas concentrations, but which is also strongly determined by the reaction of the hydrological cycle (water vapor, clouds, precipitation, evaporation, snow cover, extent of sea ice). The water cycle can intervene to reinforce or to curb this development, as many of its branches are highly dependent on temperature. Particularly important is a mechanism known as 'water vapor feedback.' One of the consequences of rising temperatures in the lower atmosphere is that this region can hold more water vapor. As we know, water vapor is a greenhouse gas, so that the initial warming is further increased because of the increased water vapor content. This is known as a 'positive feedback,' that is, a reinforcing process. Water vapor feedback is the most effective of all of the various feedback mechanisms, and, of course, it is taken into account in climate models. Since warming differs regionally and within any given year, and because the disruption to the radiation balance due to a change in greenhouse gas concentrations depends on the structure of the atmosphere as well as the season and the type of surface, an increased greenhouse effect will also result in changed values for precipitation, cloud formation, the extent of sea ice, snow cover and sea level, and for other weather extremes: in other words, in a complex global climate change. Particularly important for mankind are the potential change in the statistics of extreme weather events and a possible rise in sea levels of many meters in the long term.

2.4 The Global Carbon Balance

The greenhouse gas most important in the anthropogenic greenhouse effect is carbon dioxide, which is responsible for around 60% of the effect caused by the well mixed greenhouse gases. Methane's share is around 20%, and CFCs' just under 15%. These numbers cannot be mistaken for those of the natural greenhouse effect, for which water vapor is the dominant gas, responsible for around 60%. The lion's share of the additional carbon dioxide comes from burning fossil fuels: oil, coal and natural gas. The global emission of CO_2 is thus intimately liked with worldwide energy consumption, as the generation of energy is based primarily on burning fossil fuels. At present around three quarters of CO_2 emissions still come from the industrialized nations, although only 25% of the world's population lives there. The United States and China head the list, with around 40% of global CO_2 emissions. However, as global economic growth accelerates, the share of the tiger economies and developing countries, especially that of China and India, will increase rapidly over the next decades.

Since the start of the Industrial Revolution the concentration of CO_2 in the atmosphere has risen at a breathtaking pace. While the CO_2 content was around 280 ppm around 1800, it is already 384 ppm today. There is no doubt that humanity is responsible for this rise. A look at the past shows that today's CO_2 content is already higher than it has been for millennia. This was the conclusion reached through reconstructing fluctuations in the chemical composition of the Earth's atmosphere by analyzing the air bubbles enclosed in ice cores taken from the Antarctic. The temperatures were derived, too, on the basis of measurements of oxygen isotopes. Figure 8 shows astonishing parallels in the courses of CO_2 content and temperature over the past 800 000 years, which suggests a close connection between these two quantities. The same observation is true for temperature and the concentration of methane. Apparently there is a positive feedback between temperature and the concentration of greenhouse gases: a change in temperature leads to a change in the concentrations of greenhouse gases, which, by changing the intensity of the greenhouse effect, reinforces the initial change in temperature. Similarly, a change in the concentration of greenhouse gases results in a change in temperature, which reinforces

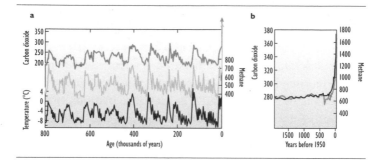

Figure 8   a) The 800,000-year records of atmospheric carbon dioxide
           (dark green, ppm) and methane (light green, ppb) from
           Antarctica, together with a temperature reconstruction
           (relative to the average of the past millennium)

           Concentrations of greenhouse gases in the modern atmosphere are highly
           anomalous with respect to natural greenhouse gas variations (present-day
           concentrations are around 380ppm for carbon dioxide and 1,800 ppb for
           methane).

           b) The carbon dioxide and methane concentrations from the
           past 2,000 years (redrawn after Brook 2008)

the initial change in the concentration of greenhouse gases.
These are the positive feedbacks that make the climate system
react extremely sensitively to relatively minor external impulses.

Figure 8 also makes clear that while CO_2 content in the atmosphere has always been subject to fluctuations, over the past 800
000 years it has typically ranged between around 200 and 300
ppm. During the last ice age around 20 000 years ago the CO_2
content was slightly under 200 ppm; during the last major interglacial, the Eemian Interglacial approximately 125 000 years
ago, it was approximately 300 ppm. Today we are thus situated in a range that is unique for humanity and apparently far
beyond the band of natural fluctuation. Hence it is extremely
improbable that the observed rise in CO_2 can be traced back to

natural causes. Moreover, modern measuring techniques leave no doubt that man is the perpetrator. Measurements of the carbon isotopes ^{12}C and ^{13}C make it possible to differentiate between carbon dioxide from natural (biogenic) sources and carbon that originates from the combustion of fossil fuels. A series of isotope measurements confirmed that the increase of carbon dioxide in the atmosphere is of anthropogenic provenance and cannot be traced back, for example, to gas released by the ocean. In addition, the carbon dioxide instrumental measurements performed simultaneously in the atmosphere and in the ocean show that carbon dioxide from the atmosphere is passed on to the ocean, such that the oceans function as a carbon dioxide sink. In the following we will concern ourselves with the global carbon cycle in order to understand what happens to the CO_2 emitted by man.

By now a global network is in place to measure CO_2. The first measurement station was set up at the Mauna Loa Observatory in Hawaii in 1958. Today the measurement network is truly worldwide: measurements are even recorded in the Antarctic. Using these measurements and many other observations, it is possible to estimate the global carbon cycle. In principle it can be said that around half of the CO_2 emitted into the atmosphere by man is taken up by the oceans and by vegetation, while the other half remains in the atmosphere. Although the carbon flows between the ocean and the atmosphere, or between land and the atmosphere, are much greater than the disruption by mankind, the anthropogenic CO_2 input is currently the dominant factor in the climate. Although the natural carbon flows are considerably larger, for the most part they are balanced: for instance, normally the ocean gives off the same amount of CO_2 to the atmosphere as it absorbs itself (Figure 9). The same is true for the flows between land and the atmosphere. In the last centuries this balance guaranteed a relatively constant atmospheric CO_2

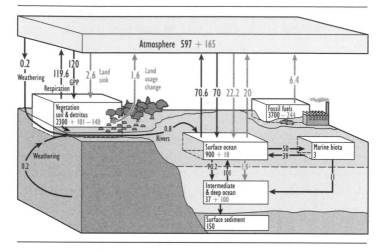

Figure 9   The global carbon cycle for the 1990s, showing the main
annual fluxes in GtC/yr: pre-industrial 'natural' fluxes in black
and 'anthropogenic' fluxes in green

Atmospheric carbon content and all cumulative fluxes since 1750 are as of late
1994 (redrawn after IPCC 2007)

concentration of around 280 ppm. Man is disturbing this equilibrium, which is why the CO_2 content of the atmosphere has risen so rapidly since the beginning of industrialization.

Table 2 depicts the global balance of atmospheric CO_2 for the period from 1990 to 1999, as recorded by measurement instruments. According to these values, global emissions from the consumption of fossil fuels (6.4 GtC/year; 1 GtC = 10^{15} gC or 1 PgC) faced an increase of 3.2 GtC/year in the atmosphere*

---

* The emissions and the atmospheric rise increased considerably during the last years, amounting to about 7.2 GtC/year and 4.1 ppm/year in the period 2000–2005, respectively. The ocean and the land

and approximately 2.2 GtC/year were absorbed by the ocean. This means that the terrestrial biosphere absorbed a net quantity of 1.0 GtC/year. However, if emissions from changed land use are also taken into account (1.6 GtC/year), including slash-and-burn land clearance in the tropical rainforests, which can be estimated from statistics on agricultural areas and assumptions about carbon content in vegetation and the soils of natural and agriculturally used areas, then sink processes contributing a total of 2.6 GtC/year would be required to balance the terrestrial biosphere. Matters of especially controversial discussion include the 'fertilization' of vegetation through the increase in atmospheric CO_2 or through the anthropogenic input of nitrogen, but also through the increased cultivation of forests, which are growing again today and thus can bind additional carbon. However, it is also possible that natural climate fluctuations (e.g. changes in ocean circulation, droughts, warming in the intermediate and higher latitudes of the Northern Hemisphere and an associated prolongation of the vegetation period) can cause variations in the atmospheric concentration of CO_2.

Emissions (fossil + cement)	6.4 ± 0.4
Atmospheric increase	3.2 ± 0.1
Ocean sink	2.2 ± 0.4
Net land sink	1.0 ± 0.6
Land use change	1.6 ± 1.1
Residual land sink	2.6 ± 1.7

Table 2 Atmospheric carbon balance 1990–99

Unit: GtC/year (Pg carbon per year)

took up 54% of the anthropogenic CO_2 in the period 2000–2007.

The global carbon cycle discussed above, based on measurements all over the world, clearly shows that the sinks are not sufficient to completely remove the CO_2 humans bring into the atmosphere, which explains the rise in CO_2 concentration since the beginning of industrialization. Studies of the other relevant materials cycles (the methane and nitrogen cycles) deliver qualitatively similar results, but will not be discussed in detail here. Overall it is clear that, in spite of the still existing uncertainties in our understanding of global materials cycles, man is definitively responsible for the increase in atmospheric greenhouse gas concentrations.

2.5 The Ozone Problem

Another climate problem is known under the title 'ozone problem' or 'ozone hole.' Ozone occurs primarily in the stratosphere, the layer at an altitude of 10–50 km. Ozone (O_3) filters out most of the UV radiation harmful for living creatures, and life was not able to move from the ocean onto dry land until after the ozone layer had formed. The ozone layer is damaged by the emission of CFCs by us humans. Stratospheric ozone must be differentiated from ground-level ozone. The latter is often generated in congested urban areas during the summer through photochemical processes caused by the emission of nitrous gases and other so-called precursor substances. Since ozone is a toxic gas, endeavors are undertaken to reduce the formation of ground-level ozone. Compulsory catalytic converters for automobiles were a first step in this direction. Strictly speaking, therefore, we are facing two different ozone problems.

Ozone is formed in the stratosphere when molecular oxygen (O_2) is split into two O atoms by ultraviolet radiation with wavelengths smaller than 242 nanometers (UV-B) (photolysis). Each of these oxygen atoms can attach itself to an O_2 molecule, thus creating ozone (O_3). Over the course of time the entire amount of atmospheric oxygen would be transformed into ozone, were there not also decomposition mechanisms for ozone in place, during which molecular oxygen is again created. This happens, first, as ozone is split by visible light with a wavelength of less than 1200 nanometers, and second, through reactions caused by collisions with oxygen atoms. For a long time it was presumed that these mechanisms provided a complete description of the fate of stratospheric ozone (Chapman cycle). However, today we know that a third category of reactions plays an important role in the destruction of ozone: catalytic reactions of ozone with radicals (HO_x, NO_x, ClO_x, BrO_x). These reactions are called catalytic because the radical initially destroys several ozone molecules, and in the end is released as an unchanged radical, so that it can run through the same reaction again with new ozone molecules. This process is repeated until the radical reacts with a molecule other than ozone and thus becomes integrated into a stable compound. Some of these radicals are of natural origin, but most originate from anthropogenic emissions. The human contribution of chlorine radicals (from CFCs) amounts to over 80%.

In any case, the formation and destruction of ozone takes place as long as the number of ozone molecules formed per unit of time is the same as the number that are destroyed.

At the point described above, the ozone concentration is in a state of 'dynamic equilibrium.' Since the reactions of forming and destroying ozone take place at different speeds in different altitudes, the equilibrium concentration depends on the altitude. The resulting vertical ozone profile exhibits especially high ozone concentrations in the stratosphere at altitudes around 15 – 30 km.

This layer is called the ozone layer, because it contains around 75% of the atmospheric ozone. Destruction of the stratospheric ozone layer would be catastrophic, for then UV radiation would be able to reach the Earth's surface practically unchecked, with incalculable consequences for terrestrial life.

The 'ozone hole' describes the rapid decrease of the concentration of ozone in the stratosphere, which has been observed to increase over the Antarctic every year since 1980. Responsible for the rapid and very effective decomposition of ozone are the chlorofluorocarbons (CFCs), which enter the atmosphere only through human agency. Essentially, the ozone problem can be regarded as independent of the anthropogenic greenhouse effect. Nevertheless, these two complexes of problems are often conflated. This is probably because of CFCs, which both cause (or at least contribute to) the anthropogenic greenhouse effect and destroy stratospheric ozone.

Important for the decomposition of ozone are the catalytic reactions of the ozone-destroying radicals, whereby the radicals containing chlorine deserve particular mention. In order for the observed fast destruction to take place, considerably more radicals must be present in the area of the ozone layer than normal. This 'enrichment' of the stratosphere with radicals takes place in two phases: in winter, chemically inactive substances that bind with catalysts ('reservoir gases') create compounds that are stable only in the dark. These 'precursor substances' are formed on the surface of stratospheric particles, undergoing what is referred to as heterogeneous reactions. One example is the reaction of gaseous chlorine nitrate ($ClONO_2$) with chlorine oxide (HCl) to form gaseous chlorine gas (Cl_2) and nitric acid (HNO_3). In this example $ClONO_2$ and HCl are the reservoir gases and Cl_2 the precursor gas. In early spring on the Southern Hemisphere (October) the precursor substances are split by the incursion of sunlight, causing the rapid release of large quantities of ozone-destructive radicals (e.g., two highly reactive chlorine atoms (Cl) are generated from chlorine gas (Cl_2)). The number of ozone molecules destroyed by catalytic decomposition reactions is then much greater than that of the molecules that can be formed. At the same time, the reactions that usually remove the radicals from the atmosphere take place much more slowly, shifting the equilibrium to extremely low ozone concentrations. A very intensive thinning of the ozone layer is observed, the 'ozone hole.' The surfaces on which the heterogeneous reactions that form the precursor substances take place are provided primarily by what are known as polar stratospheric clouds (PSC). However, these clouds are formed only at temperatures below −78°C. Even in the polar winter such temperatures are reached in the stratosphere only when the air remains devoid of any solar radiation for a long time. What is more, this air must have spent a great deal of time over the pole and may not have been mixed with warm air from mid-latitudes.

Indeed, the exchange of air is strongly impeded in the polar winter. The air over the polar region, no longer irradiated by the sun, is considerably colder than that over the mid-latitudes. This temperature difference results in air flowing toward the pole. Under the influence of the Earth's rotation, a gigantic stratospheric cyclone is generated, the polar vortex. This vortex blocks the exchange of air between the polar and mid-latitudes, so that any air that gets caught up in the vortex is, in a manner of speaking, 'trapped' over the polar region. Only in this vortex does air have enough time to cool down to make the formation of PSCs possible. Thus the formation of precursor substances from reservoir gases and the accumulation of enough precursor substances to destroy drastic amounts of ozone is only possible in a stable vortex. The polar vortex dissolves again in the spring through the gradual warming of the polar stratosphere and the corresponding change in air currents.

So far, the phenomenon of the ozone hole has occurred only over the Antarctic. From the beginning of spring there (mid-September) a rapid decrease in total ozone content can be observed, dropping to less than half the normal value. The cause is, as described above, chemical reactions that destroy ozone at certain altitude ranges of the stratosphere, sometimes completely. The ozone values then remain at a low level for six to eight weeks before the ozone content begins to rise again, attaining nearly the level of the previous year. Similar ozone losses have been observed over the Arctic in the past, but so far they have not lasted long (several days at most). Their extent, too, was far smaller than over the Antarctic, so that there is no grounds for speaking of an ozone hole here yet. The reason for this difference between the two polar regions lies essentially in the different geographies (land-sea distribution, mountains) and the correspondingly less stable meteorological conditions on the Northern Hemisphere. These create a situation in which the conditions for the occurrence of an ozone hole, and above all for an extremely stable polar vortex with very low temperatures below $-78°C$ and an altitude of between 15 and 25 km, is much rarer over the Arctic and of much shorter duration than over the Antarctic.

Even without the special conditions within a polar vortex, outside the higher latitudes stratospheric ozone decomposes more intensively. This also happens through heterogeneous reactions, which take place in this region on the surface of droplets of sulfuric acid. Although this process is not as spectacular as the ozone hole over the Antarctic, it does constitute the most important contribution to global ozone decomposition aside from gas-phase chemistry. Dynamic processes play an additional role in the ozone layer at mid-latitudes. After the polar vortex dissolves, the ozone-deficient air over the Polar Regions is transported to the lower latitudes. While this process destroys no ozone, it does reduce the density of the ozone layer at mid-latitudes. Both processes, the local destruction of ozone and the processes of exchange, result in a thinning of the ozone layer over the mid-latitudes that is slow, but constant over the years; depending on the latitude, it amounts to 2 to 7% per decade. The thinning of the global ozone layer exposes the biosphere to real dangers, as it also occurs in regions where solar irradiation is considerably more intensive than over the poles.

As we saw above, the polar stratospheric clouds play an important role in the destruction of ozone. However, since these clouds form only at temperatures below −78°C, and such low temperatures persist over several weeks only over the South Pole during the polar night and during low-exchange, very stable weather situations (polar vortex), at present the ozone hole is located only over the Antarctic. The anthropogenic greenhouse effect will warm up the lower layers of air, but actually cool the stratosphere. This carries the risk of more advantageous conditions for the creation of PSCs, through which the destruction of ozone will be promoted further. The possible recovery of the ozone layer, introduced to the international policy stage by the Montreal Protocol, may be delayed as a result.

The other possible consequence of the anthropogenic greenhouse effect is a change in stratospheric air circulation, in particular, an intensification of the Northern Hemisphere's winter polar vortex. Both stratospheric cooling and changes in circulation result in a reduction of the differences between the two pole areas. This means that the probability of another ozone hole forming over the Arctic and northern Europe increases considerably. Progressing climate change could have a further consequence for the ozone layer over Europe: it is known that tropospheric high-pressure areas lift the tropopause, the boundary surface between the troposphere and the stratosphere, and the stratosphere above it, including the ozone layer. Because the formation and decomposition of ozone is dependent on altitude, decomposition is somewhat stronger at greater altitudes, so that the ozone layer is somewhat thinner over every high-pressure area. Various climate model calculations now predict higher air pressure over Western Europe; averaged over the year, this would mean a further thinning of the ozone layer. A conclusive scientific assessment of this problem has yet to be submitted, however. For this, traditional climate models would have to be linked with elaborate chemical models. The first such combined models, which simulate the complex interactions between the physical and chemical processes, are currently under development, and first results indeed support the picture of a delayed ozone recovery in response to an enhanced greenhouse effect.

3 Climate Variability and Prediction

3.1 Why Does the Climate Fluctuate?

One of the prominent characteristics of the climate is its extreme variability, i.e. its wide range of fluctuation (see Figure 8). One year is not like the other, one decade not like the previous one. In the past, most climate changes have come about naturally. The detection of anthropogenic climate changes is complicated by their overlap with these natural climate fluctuations. Climate fluctuations are observed on a variety of time scales, from months up to millions of years. In principle we differentiate between two kinds of climate fluctuations: external and internal. While external climate fluctuations are induced by disturbances from 'outside,' internal climate fluctuations arise through interactions within or among the different climate subsystems. The best known examples of external climate fluctuations are the ice ages. These are cold periods accompanied by the abnormally large spread of ice, the causes of which include variations in the parameters of the Earth's orbit. The anthropogenic greenhouse effect also counts among the external climate fluctuations, as do climate changes caused by volcanic eruptions.

Although climate fluctuations are generally perceived as longer-term changes in the properties of the atmosphere surrounding us (e.g. air temperature or frequency of precipitation), the causes of climate fluctuations are not necessarily to be found within the atmosphere; predominantly, they can be traced back to its interactions with the inert components of the climate system (ocean, sea ice, ice sheets, biosphere). Thus, according to the concept of the 'stochastic climate model' proposed by Hasselmann in 1976, analogous to 'Brownian motion' in theoretical physics, short-period fluctuations in air temperature and wind induce long-period fluctuations in the ocean and sea ice. Changes in the surface temperature of the ocean and the extent of sea ice, in turn, lead to changes in the atmospheric processes of weather (see section 3.3). Observations show that climate fluctuations (again, corresponding to Brownian motion) become stronger as the time scale increases, such that the temperature deviations during the ice ages were greater than those in the Little Ice Age in the Middle Ages, and these, in turn were greater than the annual temperature fluctuations observed at present.

3.2 El Niño/Southern Oscillation

The strongest short-term internal climate fluctuation is the El Niño/Southern Oscillation (ENSO) phenomenon. ENSO is a classic example of the large-scale interactions between ocean and atmosphere, and is expressed through anomalies in the temperature of the ocean surface in the eastern equatorial Pacific. Although ENSO has its origins in the tropical Pacific (Fig. 10), it influences the climate far beyond its own region. The effects of the warm phases of ENSO, the El Niños, include droughts in Southeast Asia and Australia. They also cause heavy precipitation across large swaths of western South America, and bring about significant climate anomalies over North America; particularly strong events may even have effects over Europe. However, ENSO affects not only the global climate, but also the ecosystems in the Asian-Pacific area and the economies of various states, like Australia and Peru. Even the atmospheric concentration of CO_2 changes in the short term as a consequence of ENSO extremes. The simulation and forecast of ENSO and other internal climate fluctuations, such as the North Atlantic Oscillation (NAO, see section 3.3), is a welcome test of the quality of climate models. After all, realistic climate models should be able to simulate not only the mean climate, but also its range of fluctuation.

El Niño is the name used to designate the large-scale warming of the upper ocean in the eastern equatorial Pacific, which occurs every four years on average. The term 'El Niño' is Spanish for 'the Christ Child' and was coined by fishermen on the Peruvian coast back in the century before last, who gave it this seasonal name because they observed that the surface temperature of the ocean rose every year around Christmas, marking the end of the fishing season. In some years, however, the warming was so strong that the fish did not return at the usual time in late spring. These especially strong warming events typically lasted for about one year. Today only these exceptional warming events are called 'El Niño.' Analogous to this, exceptionally strong cooling events are also observed, which have been given the analogous name of 'La Niña.'

Figure 10 shows the deviation of the ocean surface temperatures from the long-term mean, as was observed during the last Super El Niño in December 1997. The large-scale character of this warming is clearly apparent: it extends across more than one quarter of the Earth's circumference in the vicinity of the equator. The warming pattern typical for El Niño exhibits its strongest temperature rises in the equatorial East Pacific, with anomalies of over 5°C off the South American coast. El Niño is also attended by changes in the surface temperature of the ocean in other regions, like, for instance, warming in the

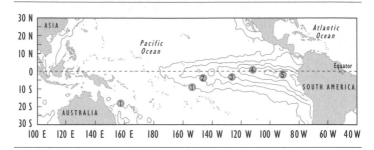

Figure 10 Anomalies in the surface temperature of the ocean observed
during El Niño in 1997, in °C

tropical Indian Ocean or cooling in the North Pacific. The latter
is induced by a change in atmospheric circulation in these areas
as a consequence of El Niño warming. Such effects are referred
to as teleconnections.

The development of El Niño is linked closely with an atmospheric phenomenon called the 'Southern Oscillation.' Because of
the close connection between El Niño and the Southern Oscillation, today one generally speaks of the El Niño/Southern Oscillation (ENSO) phenomenon. The Southern Oscillation represents
a kind of pressure swing between the Southeast Asian low pressure area and the southeastern Pacific high pressure area, and
determines the strength of the trade winds along the equator in
the Pacific. It is now known that the surface temperature of the
equatorial Pacific changes with the strength of the trade winds,
and how this occurs. Under the influence of the trade winds, cold
water wells to the surface of the ocean off the coast of South
America and along the equator in the East Pacific, causing the
ocean temperatures in this region to be much lower than those of
the western Pacific, with a temperature contrast of about 10°C.

An initial warming of the East Pacific and, connected with this, a reduced East-West temperature differential, weaken the Southern Oscillation: while air pressure climbs over the western Pacific, it drops over the East Pacific – weakening the trade winds and thus the upwelling of cold water in the East Pacific. Through this the surface temperature in this ocean region rises even higher and the temperature contrast between the East Pacific and the West Pacific is further reduced, causing additional weakening of the trade winds. This instable interaction between ocean and atmosphere ultimately culminates in an El Niño event, with the East Pacific visited by unusually high temperatures and the trade winds 'falling asleep.' Analogously, under the effects of a La Niña event, El Niño's 'cold sister,' the processes are reversed: La Niñas are characterized by a strong temperature contrast along the equator and strong trade winds.

This positive feedback between ocean and atmosphere may explain the growth and amplification of an initial disturbance, but not the oscillatory nature of the fluctuations in the equatorial Pacific. The reason for the phase reversal, that is, for swinging from an El Niño condition to a La Niña event, lies, according to one popular hypothesis, in the propagation of long oceanic waves. When the trade winds die down during an El Niño event, this initially has direct consequences for the East Pacific: the upwelling of cold water is reduced, thus promoting further warming. Yet added to this is another indirect effect: the weakened trade winds cause waves to be generated in the western Pacific, which are accompanied by an increased upwelling of cold water to the surface. However, they do not influence the temperature of the ocean surface in the western Pacific because of its deep mixed layer. These waves, which reach their maximum amplitude a few degrees north and south of the equator, first migrate westward and then bounce off the Asian

Australian continent, whereby the type of wave changes. The upwelling signal then migrates eastward along the equator with these waves. When they reach the East Pacific, where the mixed layer is shallow, the waves then cool down the temperature of the ocean surface and induce the swing to a La Niña event.

In so far ENSO can be conceived as a cycle, which thus exhibits a degree of predictability. Today climate models are capable of predicting El Niño and La Niña events, and thus the global climate changes that they entail, around six months in advance. The ENSO forecast around twenty years ago represented a breakthrough in seasonal forecasting. Today ENSO forecasts are issued operationally by various weather centers and used for planning by the governments of various states. The realistic simulation of climate phenomena like ENSO is an important prerequisite for the deployment of climate models in resolving issues concerning global climate change.

3.3 The North Atlantic Oscillation (NAO)

While ENSO primarily concerns the tropics, the climate of the North Atlantic Sector, and thus Europe and North America as well, is determined strongly by another phenomenon, the North Atlantic Oscillation (NAO) (see, e.g., Hurrell 1995). The dynamics of the NAO variability, however, are extremely complex so that a rather lengthy description is required to understand the basics of the variability in the mid-latitudes. Like the ENSO, the NAO is an internal climate fluctuation and has been known for many decades. Back in the 1920s it was described by Walker, who also discovered the Southern Oscillation. It is a sort of pressure swing between the 'Icelandic low' and the 'Azorian high.' The NAO is characterized by a dipole in the field of surface pressure

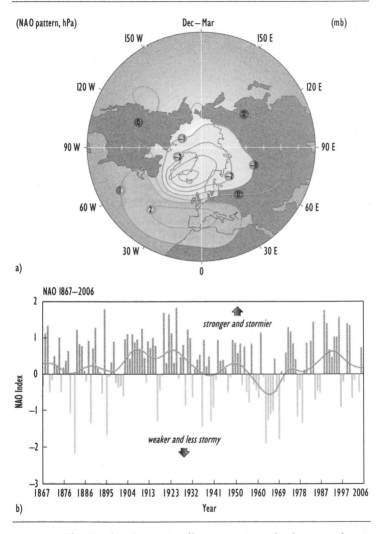

Figure 11 a) The North Atlantic Oscillation, an interplay between the
            Iceland low and the Azore high
            b) The NAO index

over the North Atlantic (Figure 11 a) and thus exerts a great influence on the intensity of the westerly winds in this region during the winter. A simple index of the NAO is the pressure differential between Iceland and Lisbon (Portugal). A high NAO index stands for an abnormally strong Icelandic low and an abnormally strong Azorian high. A low NAO index is characterized by an abnormally weak Icelandic low and an abnormally weak Azorian high.

The NAO index (Figure 11b) have shown marked interdecadal fluctuations since 1860, when barometric measurements were begun at both stations. At the beginning of the 20th century relatively high values were measured; in the 1960s the NAO index reached a minimum and then rose relatively strongly again. The rise in the final 30 years of the last century made a considerable contribution to the warming of the Northern Hemisphere in the winter, especially over Eurasia. It is not clear whether this intensification of the NAO can be traced back to anthropogenic influences or is due to natural causes. In the long term, however, we do expect an intensification of the NAO as a consequence of global warming. The amount of global warming is still quite small, however, so that even in the coming decades our climate will continue to be determined for the most part by natural fluctuations. In so far it is altogether possible that the NAO index could move into a negative phase again in the coming decades, which would increase the probability of severe winters in Northern and Central Europe and thus moderate the influence of global warming in the short term. In the long term, however, i.e. in the second half of this century, global warming will prevail over the natural fluctuations of the NAO unless mankind changes course.

Fluctuations over shorter time scales are also found in the NAO index, such as a strongly pronounced quasi-decadal oscillation with a period of around ten years during the second half of the last century. Especially striking, however, is the strong year-to-year variability. Because the time series is so short, it is difficult to demonstrate characteristic time scales. Yet it is clear that the spectrum of the NAO index exhibits some 'redness', in that the amplitudes become ever stronger as the time scales increase. This might be an indication for the fact that slow changes in ocean circulation, for instance in the thermohaline circulation, and changes in the ocean temperature associated with these, affect the atmosphere.

The changes in the NAO have consequences for the climate over the North Atlantic Sector. Both the surface temperature and the precipitation over Europe, for instance, are affected strongly by the NAO. Over the last fifty years the correlation of winter temperatures in Hamburg (Germany) with the NAO index has been about 0.8, suggesting a strong connection. The frequency of storms over the Atlantic is also closely correlated with the NAO. Phases of high NAO indexes are usually associated with mild temperatures, increased precipitation and more storms over Northern Europe. Yet the NAO is also relevant far beyond climate research, for example for the fishing industry and for the insurance and energy sectors. A summary of the consequences of the NAO is found in the 1998 report by Visbeck et al. However, it should also be mentioned in this context that individual extreme weather events can be entirely independent of the NAO. Examples of this include the abnormally heavy precipitation during the Oder flood in summer 1997 and during the Elbe flood in summer 2002, which can be traced back to the phenomenon of the Genoa cyclone (Vb-type weather regime).

Major changes in both atmospheric and oceanic circulation have been observed during the last decades, especially in the North Atlantic. The fluctuations in ocean circulation can be traced back to changes in large-scale atmospheric circulation, above all to long-period changes in the NAO.

Inversely, slow changes in the surface temperature of the North Atlantic appear to have an influence on the NAO. Thus it is important to treat at least the long-period fluctuations in a coupled context (ocean – atmosphere). As described above, the climate system comprises components with very different time scales. While the atmosphere exhibits typical time scales of several hours up to several days (the lifespan of a mid-latitudinal storm, for instance, amounts to several days), other components behave much more sluggishly. Anomalies in sea ice or in the top layer of the ocean, for instance, can linger from several weeks up to several months. Deeper layers of the ocean, which are important for processes connected with the thermohaline circulation, have adaptation times ranging from decades up to millennia. Therefore the ocean plays a particularly important role with regard to the generation of long-period fluctuations in the NAO.

Recently it has become ever more apparent that changes in the circulation of the North Atlantic and variations in atmospheric parameters are correlated. However, there are still considerable gaps in our understanding of the interactions between ocean and atmosphere in the mid-latitudes. Various hypotheses have been advanced to describe the interaction between the ocean and the atmosphere. Many aspects of the interaction between systems with different time scales can be described, like Brownian motion in statistical physics. The idea is that anomalies in the temperature of the ocean surface and sea ice are generated by the summation of many individual, statistically independent effects of the atmosphere, such as multiple high-pressure and/or low-pressure systems passing by (see section 3.1). According to this concept of a 'stochastic climate model', in which the ocean reacts passively to the atmosphere, the typical atmospheric forcing (like the heat flux between the ocean and the atmosphere, for instance) exhibits what is known as a 'white spectrum,' which means that the same amplitudes are found on every time scale. The ocean, however, reacts with a 'red' spectrum, meaning that the amplitudes of its fluctuations grow as the time scales increase. In other words: the ocean reacts selectively to the atmospheric forcing. The spectra of many long time series recording fluctuations in the surface temperature of the ocean or in surface salinity are consistent with this concept.

Through the occurrence of peaks, some climate spectra demonstrate that certain time-scale ranges are particularly selective. These can also be understood in the framework of the stochastic climate model, in which resonant interactions between ocean and atmosphere play an important role. What matters here is that the ocean can execute damped oscillations of its own, which are induced by the atmospheric weather 'noise', in the same way a swing is set in motion by a gust of wind. In this case most of the oscillations induced in the ocean exhibit a pattern on the ocean surface that resonates with the patterns of atmospheric propulsion. This generalized concept of the stochastic climate model still produces white atmospheric and red oceanic spectra, but peaks as specific frequencies of the oceanic spectrum are superimposed over the red background.

However, in some regions peaks are found in the spectra of both oceanic and atmospheric variables. These can be traced back to dynamic (mutual) interactions between ocean and atmosphere (Bjerknes 1969), in which the ocean reacts to the atmosphere and the atmosphere reacts to the ocean. The ENSO phenomenon in the tropics described above is a typical example for dynamic interactions of this kind. Similar coupled phenomena may also exist in mid-latitudes, although these typically have periods of several decades because of the relatively long response times of the oceans in these regions (Latif and Barnett 1994). Which of these scenarios is relevant for the NAO is a current object of research.

Due to insufficient observation data we are not able to describe all details of the mean ocean circulation and its fluctuations on the various time scales and to examine its interaction with the NAO more closely. Only a few long-term measurement series from the North Atlantic exist. These include the temperatures of what is known as the 'Labrador Sea water.' It originates through the cooling of surface water in the winter; as the water cools it becomes more dense, dropping to depths below 2000 m in the central Labrador Sea. This process of convection is believed to be an important motor of the deep thermohaline circulation in the North Atlantic. High temperatures point to the formation of less Labrador Sea Water, while low temperatures indicate that the formation of Labrador Sea Water has increased. The temperature of Labrador Sea Water rose from 1950, reaching a maximum in 1970 before cooling down again. This interdecadal fluctuation is accompanied by changes in the eastward Gulf Stream transport, which are inversely related to those in the temperature of the Labrador Sea. Such fluctuations in the temperature of the Labrador Sea and the Gulf Stream are coherent with the low-frequency variations in the NAO, suggesting that a coupled ocean-atmosphere interaction scenario may apply.

What these interactions look like in detail is unclear because of the poor observation data. However, it appears that there is some kind of connection between the thermohaline circulation and the NAO. Similar interdecadal variations are simulated by the global climate models. However, it has yet to be shown that the results of the models reflect actual conditions. In the models the interdecadal fluctuations are usually based on a cycle in the coupled ocean-atmosphere system. Although we do not know whether a cycle of this kind exists in reality, the model studies offer an impressive demonstration of how complex interactions between ocean and atmosphere can lead to interdecadal fluctuations in the North Atlantic and thus in the NAO as well. However, it must be mentioned that the details of such kinds of interactions can vary strongly from model to model.

No matter what kind of ocean-atmosphere interaction prevails in the real system, it certainly can be assumed that slow changes in the ocean have effects on the atmosphere. Ocean currents like the Gulf Stream transport enormous quantities of heat. Thus it is plausible that changes in the intensity of the Gulf Stream are linked directly with climate changes over the Atlantic and the adjacent land areas. Atmospheric circulation models can be used to show that there is in fact an influence of long-period fluctuations of the sea surface temperature (SST) on the NAO. The models are fed with the SSTs observed over several decades. A comparison of the observations with the simulation shows that at least the long-period parts of the NAO can be reproduced by the model, which implies a certain predictability of the NAO over long (interdecadal) time scales. However, it is also clear that the changes between the individual years cannot be reproduced accurately by the models. This suggests that the short-term changes in the NAO may be attributed above all to unpredictable, i.e. chaotic, internal atmospheric variability.

The trend observed in the NAO during the final decades of the last century repeatedly raised the question as to whether humans are influencing the NAO. Research is just beginning on this complex of issues and no conclusive assessment is possible. However, most global climate models show an intensification of the NAO should global warming continue to progress. For instance, the Hamburg climate model simulates a moderate intensification of the NAO, but also, and above all, an eastward shift in the NAO pattern, accompanied by a change in storm activity. What exactly a reaction of this kind would mean for us, or more specifically, for the agricultural or energy sector, has yet to be worked out. Our understanding of the processes that lead to the low-frequency changes in the NAO is still relatively vague. The role of the ocean, especially, has yet to be examined in detail. What we need above all is realistic (high-resolution) global ocean circulation models. Initial developments in this direction are currently being introduced at various institutes.

Also unclear is how external forcings affect the NAO, like the destruction of stratospheric ozone or a change in solar radiation. Moreover, the NAO's reaction to anomalies in ocean surface temperature, especially in the mid-latitudes, is still too little understood, which becomes clear in part through the great diversity of results from models. The role of sea ice must also be studied in greater detail. For example, it is known that an anomalous ice export from the Arctic to the North Atlantic can influence the density structure of the ocean and thus the large-scale ocean circulation, which, in turn, could affect the NAO. Finally, the role of vegetation on interdecadal time scales must also be investigated. First studies show that the interdecadal changes in precipitation over the Sahel zone can be attributed at least in part to changes in vegetation.

In contrast to ENSO, the predictability of the NAO is still largely unclear. The ENSO itself does have a degree of influence on the NAO, but it appears that the predictability of the NAO is quite limited on interannual time scales. Model simulations over longer (interdecadal) time scales suggest that there is a potential for forecasting. What is not clear over these long time scales, however, is causality. The atmospheric models do react to changes in the lower boundary conditions such as the SST, but what causes the changes in the boundary conditions and whether they are predictable themselves remains unclear. Changes in the ocean surface temperatures in the tropics and in the mid- or higher latitudes correlate with the low-frequency changes in the

NAO. The most recent results show that even the surface temperature of the Indian Ocean can have a significant influence on the NAO. Therefore it is important to work out which regions are of particular importance for the NAO.

3.4 The Influence of Volcanoes

Around two thirds of all active volcanoes are located on the Northern Hemisphere (Figure 12). Most of the active volcanoes are situated in the tropics between 10° N and 20° N. Only 18% are located between 10° S and the South Pole. The stratospheric input of gas and ashes thus influences chiefly the climate of the Northern Hemisphere.

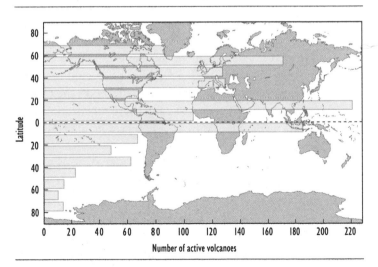

Figure 12   Number of active volcanoes as a function of geographic latitude (source: Schmincke 2004)

Large volcanic eruptions have always been linked with weather and climate anomalies. Particularly strong volcanic eruptions cause the amount of sulfuric acid aerosol in the lower stratosphere to rise by one to two orders of magnitude. This sulfuric acid aerosol is formed through the oxidation of magmatic sulfurous gases (above all SO_2 and H_2S) after their transport to the stratosphere. The late 19th and early 20th century was a phase of particularly high volcanic activity; after a lull of 50 years during which no especially strong eruptions occurred, the eruption of the Agung Volcano on Bali in 1963 ushered in a new series. Volcanoes appear to have become more active in the last two decades of the 20th century, releasing correspondingly more volcanic gases into the atmosphere. The effects of volcanic disturbances subside very slowly because aerosol droplets remain in the stratosphere so long. Since the stratosphere lacks rainout and washout processes, the most effective cleansing processes in the troposphere, gravity is practically the only force that can remove them from this layer of the atmosphere, but the droplets are so small that gravitational effects take time to kick in. The mass of aerosol decreases with a half-life of around one year, so that a strong eruption exerts obvious effects on the climate for around two years. Therefore the relatively short-term effects volcanoes have on the climate are clearly subordinate to the effect of anthropogenic influences.

Satellite measurements deployed to observe the ozone layer, discovered large quantities of SO_2 after the eruption of El Chichón (1982); only then was it discovered that the volcanic aerosol in the lower stratosphere did not comprise solid ash particles, but rather sulfuric acid droplets with radii of 0.1 to 0.5 µm. These small droplets can interact especially intensively with radiation by partially backscattering visible light and absorbing radiation in the near-infrared and long-wave ranges of the spectrum. One of the consequences of this is that less solar radiation penetrates through to the Earth's surface, with cooling effects for the lower atmosphere: in the cases of the Krakatoa eruption in 1883 and Pinatubo in 1991, temperatures fell by an average of around 0.1–0.2°C worldwide. On the other hand, the absorption of radiation by volcanic aerosol has a considerable warming effect on the stratosphere, with about 1–2°C in its lower part. The aerosol droplets also have chemical effects: by facilitating the activation of chlorine compounds, they ultimately support the destruction of ozone. The two most important volcanic gases in terms of mass, water vapor (H_2O) and carbon dioxide (CO_2), are relevant for the climate only over extremely long geological time scales, since the amounts emitted are negligible in comparison to their concentration in the atmosphere. Volcanic ashes fall out of the atmosphere quickly, exerting only a short-term influence on radiation transport and the dynamics of the atmosphere. The climate effect of volcanoes is thus due primarily to the sulfurous gases (SO_2 and H_2S), which are oxidized into gaseous sulfuric acid (H_2SO_4) when they reach the stratosphere. The increased concentration of sulfuric acid increases the stratospheric background aerosol, first through gas-to-particle conversion, which increases the number of particles; and second through the condensation of H_2SO_4 and H_2O on existing particles, thereby increasing their radii.

Particles of volcanic aerosol in the stratosphere typically have effective radii in the same range as the wavelengths of visible sunlight. Therefore they are especially effectual at scattering sunlight. In the case of large eruptions like El Chichón in 1982 and Pinatubo in 1991, direct solar radiation is reduced on the order of 100 W/m². Diffuse radiation increases by nearly the same amount, making the daytime sky appear milky white. The difference between reduced direct radiation and increased diffuse radiation on the Earth's surface amounts to only 1–10 W/m², but results in a cooling effect. An unusually strong volcanic eruption, of the Tambora in Indonesia, took place in the summer of 1815. The following year, 1816, went down in the history of North America and Europe as the year without a summer. On the east coast of the U.S. blizzards in the middle of summer led to catastrophic crop failures. The atmospheric effects of a volcanic disturbance depend on the geographic latitude at which they occur. Tropical volcanoes can affect the global climate system, because the eruption cloud can spread out over both hemispheres. Eruptions in the mid- to high-latitudes generally have effects felt only on their own hemisphere.

In addition to scattering visible sunlight, the absorption of radiation also plays an important role. In the upper part of the aerosol cloud, near-infrared solar radiation is absorbed. This effect outweighs the increase in long-wave radiation emitted by the aerosol, producing a clear warming effect in the lower stratosphere. The increased absorption of long-wave terrestrial radiation in the lower part of the aerosol cloud makes a more or less equal contribution to this warming effect. The stratospheric aerosol layer warms up most where surface temperatures are highest and solar radiation strongest: in the tropics. The resulting temperature difference between low and high latitudes results in circulation anomalies, not only in the stratosphere, but in the troposphere as well; in the Northern Hemisphere's winter, these anomalies can completely mask the pure radiation effect of major tropical volcanic eruptions.

Volcanic aerosols also have a strong influence on chemical processes. The most important effect is on stratospheric ozone. The reactions that form and destroy ozone are dependent on ultraviolet radiation, as well as the temperature and presence of surfaces on which heterogeneous chemical reactions can take place (see section 2.5). All of these parameters are influenced by volcanoes. The warming of the stratospheric layer that conveys aerosol elevates the isentropic surfaces, so that the ozone transported on them makes its way into higher atmospheric layers. The higher energy density of the solar radiation there leads to increased photodissociation, causing a drop in the equilibrium concentration of ozone, and thus to effective ozone depletion. The sunlight backscattered and repeatedly re-scattered by aerosol can heighten this ozone-depleting effect.

The heterogeneous chemistry that has contributed to the creation of the Antarctic ozone hole takes place on the surface of elements of polar stratosphere clouds (PSCs), which are composed of nitric acid and water. PSCs form only at extremely low temperatures and are thus rarer over the Arctic. In heterogeneous reactions the anthropogenic chlorine from CFCs is activated in the stratosphere, paving the way for rapid destruction of ozone once the polar night is over. Through the intensification of the Arctic polar vortex after volcanic eruptions, and the extremely low temperatures associated with this development, more PSCs form and ozone depletion is intensified in the spring. However, quite similar reactions can also take place on the volcanic aerosols composed of water and sulfuric acid. This process is then no longer restricted to the extremely cold polar areas, but effective globally, all year round. Thus after the eruption of Pinatubo in 1991 a reduction of 2% in total ozone levels was measured in the tropics, and of 7% in the mid-latitudes. Ozone depletion was even stronger inside the aerosol cloud, reaching 20 to 30% in the mid-latitudes of the Northern Hemisphere. Under natural conditions, i.e., without anthropogenic CFCs, an increase in stratospheric ozone would have been expected.

Major volcanic eruptions, in which several million tons of sulfur dioxide and/or hydrogen sulfide are emitted into the stratosphere, are thus generally capable of influencing the climate for a period of at least one to two years. The absorption and scattering of solar radiation and/or terrestrial thermal radiation due to volcanic aerosol in the stratosphere produce anomalies in the global radiation balance. These can result in changes in the general circulation of the atmosphere, and their consequences for temperature and precipitation may be even greater than the pure radiation effects. Humanity's amplification of volcanic aerosols' effects on stratospheric ozone is evident. Only since humans began releasing CFCs, creating a chlorine reservoir in the stratosphere (see section 2.5), has the sulfuric acid aerosol of volcanic origin produced the effect of ozone depletion.

3.5 The Ice Age Cycles

Climate changes on the time scale of many millennia (see Figure 8), such as the emergence and decay of ice ages, are based in part on changes in the Earth's orbit around the sun, as Milankovitch proposed in the 1930s. He used the Newtonian laws of celestial mechanics to develop his theory. Because the variations in the parameters of the Earth's orbit result in regional and in part also global changes in the solar irradiation incident at the top of the Earth's atmosphere, he posited that the resulting long-period climate fluctuations must be externally induced. For a long time the Milankovitch Theory was not accepted, but today it is generally recognized.

Three phenomena are relevant in the Milankovitch Theory. First, within a period of ca. 100 000 years the Earth's orbit around the sun changes from an ellipse to nearly a circle, and then back to an ellipse (eccentricity). These periods of 100 000 years can be detected most clearly in reconstructions of the paleoclimate, for instance in the reconstruction of the CO_2 content of the atmosphere and of the Antarctic temperature (see Figure 8). These cycles change the average annual global solar irradiation by up to 0.7 W/m². Second, the tilt of the Earth's axis changes obliquity with a period of approximately 41 000 years, varying between about 22° and 24.5°. Today the tilt of the Earth's axis amounts to just about 23.5°. The changes in the tilt of the Earth's axis influence the seasonal amount of solar irradiation entering at the top of the atmosphere. At high latitudes this can amount to an average of up to 17 W/m² annually. Third, the Earth is not a perfect sphere, but has a 'bulge' at the equator, which causes the Earth to 'wobble' such that its axis describes a circle in space (precession). This effect determines the season in which we are closest to the sun, and exhibits periods of 19 000 to 23 000 years. Today the Northern Hemisphere is closest to the sun in winter and furthest away in summer. Yet 11 000 years ago the situation was reversed. The annual average shows no net change in insolation, but the strength of the seasonal cycles does change.

The three cycles proposed by Milankovitch, with periods of 100 000, 41 000 and 19/23 000 years, can indeed be found in paleoclimate reconstructions of temperature. This is considered to be the most important proof for the validity of the Milankovitch Theory. However, other processes are generally responsible for converting what are initially relatively minor disturbances in solar irradiation into massive climate changes. For instance, eccentricity is connected with an average change in global irradiation of only approximately 0.7 W/m², but extremely strong climate changes are observed with a period of 100 000 years. The change in the parameters of the Earth's orbit alone thus cannot be solely responsible for the emergence of ice age cycles. They may be the impulse, but other processes, such as changes in the biogeochemical cycles and an associated shift in the concentration of atmospheric greenhouse gases, have an amplifying effect. Low CO_2 concentrations, as well as minimal methane content during the ice ages, amplify the initial cooling due to a weakened greenhouse effect. As such, the carbon dioxide content of the atmosphere during the height of the last ice age around 20 000 years ago was considerably lower than it is today (see Figure 8). Moreover, a higher degree of ice or snow cover resulting from this cooling will contribute to further cooling by increasing the reflection of solar radiation into space. This feedback mechanism is called ice-albedo feedback. A further positive feedback mechanism, water vapor feedback, was described above in our discussion of the anthropogenic greenhouse effect. There is also a plethora of negative feedback mechanisms, of course, which exert cushioning effects and ensure that the climate system is not thrown completely out of joint. Realistic climate models must take into consideration this abundance of positive and negative feedbacks. Simple models that contain only a few of these feedback mechanisms cannot do justice to the complex nature of climate variability.

Natural climate fluctuations on time scales of up to several centuries are generated predominantly by the large-scale interactions between ocean and atmosphere, that is, internally generated effects. On the longer time scales of millennia they are caused primarily by external influences like the variation in the parameters of the Earth's orbit, as described above. Reliable climate models utilized to simulate and detect the anthropogenic greenhouse effect therefore must be especially capable of depicting the interaction between ocean and atmosphere correctly, and of simulating the resulting climate fluctuations realistically.

Above and beyond this the models also must be able to describe the externally driven climate fluctuations that occur in response to anomalous stimuli like changed solar irradiation.

Today's climate models are able to simulate the natural variability of the climate. This is true for both the internal and the external fluctuations. As such, phenomena like the 'green Sahara' can be simulated using climate models. Around 6000 years ago the Sahara still had relatively lush vegetation, but then it became a desert almost spontaneously. The turnaround took place because of a gradual change in solar irradiation as a consequence of precession, which resulted in a decrease in summer insolation on the Northern Hemisphere and in the monsoon rains in West Africa. By entering these changes in the solar forcing into the model the desertification of the Sahara can be simulated realistically. Processes of vegetation dynamics had a positive feedback effect in this transformation. In a similar way it is possible to simulate the climate of the last millennium and quantify the contribution to the changes observed made by certain processes.

3.6 Abrupt Climate Changes

Yet, the climate strongly changes not only on the very long Milankovitch time scales of many millennia, but also over much shorter periods of decades and centuries. This is known today through the analysis of ice cores, for instance from the Greenland ice sheet. The ice cores drilled in Greenland, above all the European and American cores on the summit of the ice sheet completed in 1992 and 1993 (GRIP and GISP2), have supplied important information about the climate history of the past 100 000 years. Especially striking is the large range of climate variability during the past glacial period and the extremely rapid climate shifts. Furthermore, the switch from the last ice age to today's interglacial age in the period 20 000 to 10 000 years ago took place not gradually, but in a sort of roller coaster ride. The ice cores showed extremely fast climate shifts, which previously had been held to be impossible. This necessitated a search for new causes that could explain such rapid changes. The climate did not stabilize until around 8000 years ago, and it is this unusual stability of the climate that encouraged the development of mankind.

The kilometers-thick ice of Greenland is made up of many thousands of layers of snow, which accumulate year on year and gradually compress the older snow below, turning it into ice. Through state-of-the-art analysis methods, especially the measurement of certain oxygen isotopes, the ice cores allow the history of Greenland's climate to be reconstructed. Yet these results cannot simply be applied to the entire globe without further ado; reconstructing global climate changes requires many ice cores and cores of ocean sediment from different locations. We know today, chiefly through climate simulation models, that changes in the circulation of the Atlantic can take place very rapidly and thus could have triggered the rapid shifts in climate recorded in the ice of Greenland, at least on the regional scale (see section 4.5).

The data from Greenland possess a relatively high temporal resolution – individual years can be recognized and counted like tree rings – and are well suited to identify abrupt and very strong climate swings. During the last glacial period, on several occasions, the temperatures in Greenland rose by 8–10°C within just a few years, not returning to the normal, cold ice age level again until centuries later. These climate shifts are known as Dansgaard-Oeschger events. More than twenty of such climate shifts were counted during the last glacial period, which lasted around 100 000 years. In addition to the 'warm' Dansgaard-Oeschger events there are also extremely 'cold' phases, known as Heinrich events.

In addition to the ice cores, sediment cores of similarly high resolution were taken from the Atlantic. The layers of mud from the deep sea, some of them drilled in subtropical latitudes thousands of kilometers away from Greenland and analyzed using entirely different methods, recorded the same climate events as Greenland's ice. The Dansgaard-Oeschger and Heinrich events were therefore climate shifts that were not geographically restricted to Greenland. One possible explanation for the rapid climate shifts is a change in the thermohaline circulation (see section 1.2). While the Dansgaard-Oeschger events presumably can be traced back to an internal fluctuation in the thermohaline circulation, it is highly probably that the Heinrich events hark back to gigantic ice masses sliding into the sea from what was then the North American ice sheet. This resulted in a massive influx of freshwater into the North Atlantic and, as a consequence, weakened the thermohaline circulation. This circulation change can arise within just a few years or decades and thus offers an explanation for the rapid climate shift.

The fact that the climate swings were observed in many regions of the Earth also suggests that the cause was the thermohaline circulation, since its changes are felt worldwide. The thermohaline circulation possesses different states. For instance, it can be disrupted in the area of the North Atlantic, which is presumably what happened during the Heinrich events. Inversely, however, it can also extend extremely far to the north, which might offer an explanation for the Dansgaard-Oeschger events. Around 8000 years ago, however, this turbulence in the climate suddenly ceased and since then we have enjoyed an astonishingly stable climate. This calm presumably is also connected with the fact that the thermohaline circulation is in a very stable mode today.

3.7 The Predictability of the Climate

The atmosphere and its shifting weather is a textbook example of a chaotic system with quite limited predictability. From the groundbreaking work 1963 by Lorenz – one of the founders of chaos theory – we know that the theoretical limit for weather forecasts is around two weeks (see section 4.4). In chaotic systems like the atmosphere, very small errors in the starting conditions or in the formulation of the model grow exponentially, diminishing the quality of weather predictions within just a few days. Since weather forecasts are, mathematically speaking, an initial value problem, and we can never determine the starting (initial) condition exactly, we will not be able to extend this theoretical limit of weather forecasting of around two weeks by any decisive length in the future.

Besides this theoretical limit of weather predictability, however, there is still predictability in the atmosphere, which is known as seasonal predictability. Seasonal prediction, in contrast to weather forecasting, is not about forecasting individual weather elements like the path of a storm, but focuses on predicting the weather averaged over a certain period, that is: the climate. This could mean the average surface temperature, for instance, or precipitation over a certain region and season. Seasonal predictability is rooted in slow changes in the boundary conditions, like, for example, anomalies in the temperature of the ocean surface or the extent of sea ice; these, in turn, can be induced by changes in ocean currents like the Gulf Stream. Anomalies of this kind can influence the statistics of weather, so that the climate is predictable to the extent that the anomalies in the boundary conditions are predictable. Chaotic systems like the atmosphere are thus predictable under certain conditions, even on time scales of seasons or possibly years or decades.

This is also the basis for simulations concerning anthropogenic climate change, in which the changes in the chemical composition of the atmosphere (another boundary condition) play a decisive role. Global change predictions thus aim not to predict the weather itself, say on a specific day in the year 2100, but rather the statistics of the weather in the case of elevated greenhouse gas concentrations.

Once again, this raises the question as to the extent to which chaotic systems like climate react to human influences and whether they can be calculated at all (see section 4.4). The best method is to compare the influence of humans on the climate with a loaded die. The die is 'loaded' through our activities that raise the temperature of the Earth by emitting certain gases like carbon dioxide. This leads to more extreme weather events: in our analogy, the loaded die will come up six more often. However, we cannot say when the next six will come, for the sequence of the numbers remains random. Extreme weather occurs in a similar way: while we can calculate statistically that extreme weather events will become more common as a consequence of global warming, we do not know when precisely this extreme weather will occur.

The example of the loaded die also illustrates how the fact that a certain event has been observed once before allows no inferences at all about the properties of the die: the die already rolled sixes even before it was loaded. Translated to the weather, this means: the simple fact that heavy flooding or a long drought has been observed before does not allow us to conclude that weather statistics have not changed. In fact, observations of the last hundred years show that extreme weather events are becoming more frequent worldwide, as predicted by the climate models. And it is precisely this accumulation of extreme weather events that can be attributed to global warming. The analogy of the loaded die further illustrates that it is not possible to attribute individual weather extremes, like Hurricane Katrina of 2005, to global warming, just as it is not possible to attribute a certain six to the loading of the die. The statistics of extreme weather must always be considered over a longer period to illuminate the connection between extreme weather events and global warming.

As long as the die is not loaded too heavily, some rolls will still produce the other numbers. As an example of this, take the severe winter Germany experienced in 2005/2006. Global warming is still quite small, so that we cannot expect to have only mild winters from now on. The one on the die, or a severe winter, is still possible. We have indeed observed many mild winters in the last decades, which certainly emphasizes man's influence on the climate in the form of global warming. However, in the coming decades we must also be prepared for severe winters, as global warming is still so minor that natural phenomena, like a negative phase of the NAO, for example, can influence on our climate events. Nevertheless the probability of the occurrence of severe winters will decrease ever further, in keeping with the magnitude by which global warming increases.

4 Climate Modeling

4.1 Climate Models

The climate system, consisting of the various climate subsystems (atmosphere, ocean, cryosphere etc., see Figure 1), is a physical system and thus subject to the laws of physics. The three most important laws for the climate are the conservation laws of mass, momentum and energy. The physical laws can be represented in the form of mathematical equations. The temporal evolution of the climate is described in full by these equations when initial and boundary values are stipulated. The boundary values in these equations can also vary over time, like, for instance, the concentration of carbon dioxide or other greenhouse gases.

Yet these equations are so complex that their solution is not possible using tools like the ones applied to simple quadratic equations. Mathematically speaking, an analytical solution does not exist. However, there is also the possibility of using the methods of numerical mathematics to calculate approximate solutions. The Earth's surface is overlaid with a grid, allowing the equations to be 'discretized;' in other words, the equations, which are continuous themselves, are converted into what are called 'finite difference equations.' The equations are formulated at each point on the grid, producing many hundreds of thousands of linked finite difference equations to be applied to today's typical grid widths of a few hundred kilometers. Equation systems of such high dimension can only be solved with the help of powerful computers. The accuracy of the solutions generally depends on the fineness of the selected calculation grid, i.e., how far away from each other adjacent grid points are located.

Since these numerical methods are quite computation intensive even on powerful computers, even the supercomputers available today present limits in our choice of resolution: using a calculation grid of 1 km, for instance, is not possible because the computations would take much too long. Moreover, processes that take place on a characteristic scale smaller than the grid width have to be 'parameterized.' Physical processes like the formation of clouds, for instance, which takes place on quite a small scale, cannot be simulated explicitly; rather, they must be represented (parameterized) on the basis of the information available at the grid points (temperature, humidity, etc.). Convection in both the ocean and atmosphere, which is of singular importance in driving the general circulation in the ocean and atmosphere, is one of these processes. The totality of all mathematical equations and physical parameterizations is called a climate model. The main sources of error in climate models, besides the numerical methods, include such physical parameterizations. Yet comparison with the observations shows that the climate models can indeed simulate past climate states, today's climate and its range of fluctuation realistically. Therefore they should also be able to calculate the future climate reliably.

Climate models require certain boundary conditions as inputs, such as the solar radiation incident at the top of the Earth's atmosphere, the chemical composition of the Earth's atmosphere and the land-sea distribution. When these and other parameters are given, the climate models then simulate the three-dimensional wind distribution, the ocean currents, the temperature distribution in the atmosphere and ocean, the sea ice cover and many other climate parameters. By feeding the model information about a trend, say, the development of the atmospheric concentration of greenhouse gases over time, the system's reaction to these changing boundary conditions can be calculated, as is the case for calculations on global climate change (see Figure 24). Yet climate models can also be used to examine the dynamics of internal climate fluctuations, like the causes of the El Niño phenomenon or its predictability. El Niño forecasts are already a regular task performed at various weather forecasting centers.

4.2 Clouds and Precipitation

As described above, many physical processes depicted in the climate models lack the desired precision, as they possess typical length scales smaller than the size of the mesh in the computational grid. These include the formation of clouds and precipitation. The example of the formation of clouds and precipitation is to be used here to illustrate the complexity of such processes. The air of our atmosphere always contains water vapor, but the density of this gas can only be increased up to a certain saturation value at any given temperature. This maximum possible water vapor content of the air increases disproportionately with temperature. Inversely, a volume of air can always be cooled down to a temperature at which condensation begins, to the point where water passes from the gaseous phase into the liquid phase. This temperature is called the dew point. When cooling continues, the surplus water vapor accumulates either on solid surfaces (dew) or on cloud condensation nuclei, whereby droplets are formed. Aerosols, the very small atmospheric particles which are almost always present in abundance in the air from all kinds of sources, serve as the nuclei for cloud condensation.

When a parcel of air rises, it cools down: the water vapor it contains condenses as soon as the dew point is reached. Sunlight is scattered by the droplets this produces, creating a diffuse white light visible to the human eye as a cloud. A good example of this process is the ascension of warmed parcels of air from the ground in the summer, which results in fair-weather clouds when the lift is low, but continues to higher altitudes in cases of high uplift, producing thunderstorms with thick clouds, heavy precipitation, and even hail.

The counterclockwise rotation of storms in our northern latitudes transports warm, humid, subtropical air to the northeast. Because this air has such a low density, it is lifted when it glides into the colder, northern air. The air then cools, causing cloud formation on the warm fronts of the mid-latitude storms. Inversely, the cold air on the west side of the storms from the north pushes itself under the warmer subtropical air, lifting it above the condensation level, which then causes clouds on the cold front. Condensation and cloud formation also result when air cools because it is forced up to higher altitudes to surmount an obstacle, a phenomenon that is observed often on the front side of mountains.

Since their tiny diameter of just about 1/100 mm makes cloud droplets especially light, they float nearly freely. Through coincidental collisions some droplets gradually become larger than the others. They begin to fall, first slowly, and then ever more quickly as they gather additional droplets and become heavier. When the cloud is thick enough for its droplets to become larger than around 1/10 mm, despite continued evaporation as they fall, the droplets survive their descent through the cloud and reach the air below and the surface. However, this process of the creation of 'warm rain' explains only the usually weak drizzle from near-surface clouds in our climate zone, or the strong rain in the warm tropics from clouds at altitudes below 5 km.

More important for the formation of strong precipitation in the mid-latitudes is the path via the ice phase. Although temperatures at the cloud level often lie below 0°C, the cloud droplets themselves often remain supercooled in the liquid state. In this case, quite minor incidents (e.g. collisions or suitable ice nuclei) are sufficient to cause individual drops to freeze spontaneously. But since water vapor condenses more easily on ice particles than on water drops, these ice particles grow more quickly than the surrounding drops and begin falling earlier. They then increase in size by gathering other droplets, becoming so large that they reach the Earth's surface. Because these ice particles reach ever warmer layers on their way down, they frequently melt and arrive on the ground as rain. If, however, they become so large in a particularly thick cloud that they can no longer melt before reaching the ground, snow pellets are the result. Sometimes updrafts in storm clouds are so intensive that ice particles whose surface has melted are ripped upward again and freeze anew. At some stage they begin falling again, partially melt again, and continue to grow by gathering new droplets. This process can be repeated several times and ultimately results in hail.

An interesting question in connection with global climate change concerns the reaction of clouds to global warming. Climate model simulations show that the effect of clouds is generally to dampen the anthropogenic greenhouse effect, and that clouds thus function as a negative feedback mechanism. One consequence of higher water vapor content in a warmer atmosphere is increased cloud cover over the Earth, causing the increased reflection of solar irradiation and thus a significant cooling effect. However, this effect is clearly smaller than that of the additional greenhouse warming caused by man, so that while the clouding effect may mitigate global warming, it can by no means compensate for it entirely. What is more, it must be assumed that in a warmer atmosphere, strong precipitation would increase over many land regions. At least this is what most climate models simulate. For the record, however, the description of clouds and precipitation remains one of the greatest uncertainties in climate models today.

4.3 The Role of Condensation Trails

Condensation trails are artificial ice clouds caused by aircraft at an altitude of around 10–13 km, where the surrounding temperature of 40–70°C below zero is extremely low. The maximum moisture content of the atmosphere (beyond which condensation and cloud formation occur) becomes extremely small as temperatures decrease. The combustion of 1 kg of kerosene in an aircraft turbine generates 1.25 kg of water vapor and 3 kg of CO_2, as well as nitrogen oxides and soot. It should be mentioned that the climate effect through the emission of CO_2 by airplanes themselves is negligible. While the additional amount of water vapor near the Earth's surface is insignificant in comparison to the natural quantity, at cruising altitudes it often results in condensation directly behind the aircraft. The probability of condensation trails forming increases as the temperature of the surrounding air decreases. Yet if the relative humidity of the surrounding air is low at flight altitude, condensation trails cannot remain long after formation, as they evaporate when they mix with the surrounding air. If the surrounding air humidity lies above the value necessary for the formation of ice, however, the ice particles that form in the condensation trail can survive for long periods. Such ice clouds spread, cover large areas and can remain visible for hours or even days.

The visibility of condensation trails, but also their influence on the climate, depends essentially on their optical properties. These are determined by the number, size and shape of the ice particles. Two opposing effects play a role in this: in the solar (short wave) range of the spectrum, the ice clouds have a cooling effect because they reflect solar radiation into space. In the thermal (long wave) range of the spectrum, however, ice clouds have a warming effect. The resulting net effect is initially unclear, since the competition between the backscatter of solar energy and reduced infrared thermal irradiation depends on various factors. Because of their small optical thickness, however, condensation trails tend to have a net warming effect.

The increase of cloud coverage with ice clouds from condensation trails is particularly important for a potential influence on the climate. The evaluation of satellite and other observation data yields a regional, sporadically significant increase in cloud coverage through long-lasting ice clouds originating from condensation trails. Over central Europe and the trans-Atlantic flight routes, an average coverage by such clouds is found to be around 0.5%; outside the main flight routes the increases are even smaller. This value compares with an average cloud coverage by natural ice clouds of around 20%. In climate simulations, this currently minor increase in cloud cover produces only an insignificant contribution of 0.05°C. However, the simulations also show that if the coverage increases by an additional factor of ten, a significant contribution to climate changes certainly can be expected. With the rapid increase of flight traffic predicted, the influence of flight traffic on the climate is therefore expected to become stronger in the next decades.

In addition to the effect of condensation trails themselves, it is presumed that the additional condensation nuclei continue to have an influence on the greenhouse effect after the condensation trails dissolve through evaporation. So the number of ice nuclei in the general altitude of the tropopause could increase strongly enough to facilitate the subsequent formation of additional ice clouds (cirri). But since these can no longer be directly attributed to aircraft emissions, they elude such investigation. Yet the increased observation of 'natural' cirri in the past decades could point to such an influence.

At this juncture let us elaborate on what are known as 'chemtrails.' Ever since the journal *Raum & Zeit* (127/2004) published the article 'Die Zerstörung des Himmels' ('The Destruction of the Sky') many concerned citizens have asked whether there is any truth to the stories about chemtrails, which supposedly contain chemicals introduced by aircraft into the atmosphere. The article maintains, among other things, that military and civilian aircraft working on secret U.S. projects emit aluminum and barium compounds into the atmosphere, from which these chemtrails then develop, similar to the formation of condensation trails (the term chemtrails emerged in analogy to '*contrails*' for condensation trails). These emissions, the article asserts, are intended to counter the warming induced by the anthropogenic greenhouse effect.

There is no scientific evidence for the introduction of aluminum compounds into the atmosphere or the formation of so-called 'chemtrails.' According to Germany's Federal Environment Agency, the phenomena described are not even known to the German Aerospace Center (DLR). For many years scientists at the DLR's Institute for Atmospheric Physics have carried out studies on the effect of air traffic emissions on the atmosphere – including many measurements of gaseous and particle emissions by commercial airliners. If there were such a thing as chemtrails, the DLR should have information about them; however, the measurements contain no indication of such a phenomenon. The *Deutsche Flugsicherung GmbH* (DFS) confirmed that air traffic control monitors have not observed any conspicuous flight movements that lend credibility to the accounts of chemtrail formation.

What is more, the German Weather Service reported that no peculiarities in the observation data could be found that might indicate anomalous forms of condensation trails. Not even Germany's Federal Ministry of Defense reports any such findings. The headquarters of the U.S. Air Force in Europe declares that the projects described at the U.S. Air Force neither exist nor ever existed. Not even the World Health Organization of the United Nations (WHO) has found any evidence for the existence of chemtrails.

As mentioned above, the formation of cirrus clouds from condensation trails makes a particularly significant contribution to air traffic's climate effect. Condensation trails and cirrus warm the climate. Therefore it would be counterproductive to use additional cirri or cirrus-like clouds as a means of combating climate warming attributed to the anthropogenic emissions of greenhouse gases.

4.4 The Lorenz Model

Several fundamental characteristics of the atmosphere can be illustrated using a simplified model that was developed by Lorenz in 1963. The Lorenz model is non-linear and can be used to establish a new correlation between concepts like weather, climate and climate change. Lorenz differentiates between forecasts of the first kind and forecasts of the second kind: weather forecasts are forecasts of the first kind; forecasts on global change are forecasts of the second. Forecasts of the first kind are initial value problems. With knowledge of today's atmospheric (and, where applicable, oceanic) conditions and the physical laws governing motion, we attempt to predict the weather of tomorrow, the next week, or the weather averaged over the next season. In contrast to this, the forecasts of the second kind do not depend on the initial conditions. They ask how the statistical characteristics of the atmosphere (for example, the global temperature of the Earth averaged over the year, or the number of blizzards to be expected over Germany, or hurricanes over the Atlantic) change when a certain parameter changes, like, for instance, the CO_2 concentration of the atmosphere.

Lorenz discovered that forecasts of the first kind, although deterministic, cannot be extended infinitely into the future. For suitable parameter values of his model equations (primed variables designate the temporal derivative of these variables),

$$X' = -\sigma X + \sigma Y$$
$$Y' = -XZ + rX - Y \qquad\qquad (1)$$
$$Z' = XY - bZ$$

which possess two fundamental characteristics, instability and non-linearity, giving rise to a phenomenon known today as 'sensitivity on initial conditions,' or 'the butterfly effect.' Within a very short time, even the most minor disturbances in the initial conditions lead to a divergence of the model trajectories. Such a disturbance could be as minor as the stroke of a butterfly's wing: hence the name 'butterfly effect.' Weather predictions are thus limited in principle. Errors in the determination of the initial conditions and errors in the formulation of the model accrue rapidly, making every forecast worthless after a certain period, on average after a maximum of 14 days. The detailed study of this phenomenon ultimately resulted in the formulation of chaos theory. As such the atmosphere is the epitome of a chaotic system.

The Lorenz model contains many qualitative similarities with actual large-scale atmospheric circulation. For example, it simulates 'regime behavior,' meaning that the system tends to persevere in certain states. Such regimes are also observed, including the west wind phases that last for weeks in the winter, during which it is relatively humid and warm in northwestern Europe, and the stable high pressure situations in winter marked by cold, sunny weather. In contrast to this, west wind situations that last long during the summer are characterized by relatively cold and rainy weather in this region, and persistent high pressure phases by warm and sunny conditions. The transitions between such regimes take place randomly. For certain values of the parameters r, σ, and b the evolution of the state vector (X, Y, Z) yields the famous Lorenz attractor with the two so-called butterfly wings (Figure 13).

There are two characteristic time scales in the Lorenz model. The first describes the evolution of the system around the (weakly) unstable fixed point at the center of each of the two butterfly wings. The second describes the duration of a particular sojourn within one of the two butterfly wings. These two time scales can be clearly recognized in Figure 14 which shows the evolution of variable X in two simulations. Figure 14 also shows what we mean by chaos. When a simulation is repeated with the same parameters but slightly altered initial conditions, the system develops completely differently after quite a short time. This sensitivity with regard to initial conditions is what distinguishes a chaotic system.

Typical lengths for remaining in a regime lie in the range of a few weeks, while the shorter time scales on which the regime change takes place, known as 'synoptic,' amount to just a few days.

While Lorenz originally introduced his model to study

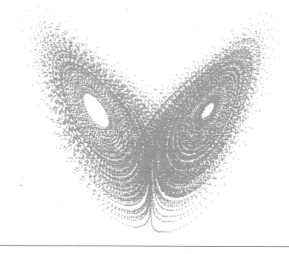

Figure 13   The Lorenz attractor in the X–Y plane. Calculated from the
above equations using the parameters r = 28, σ = 10, b = 8/3
The Z axis shows the vertical direction. From Palmer (1993)

predictability of the first kind, below we will use the model to
investigate more closely forecasts of the second kind. To do this
we introduce an external forcing vector F, which has the components F_x, F_y and F_z.

$$X' = -\sigma X + \sigma Y + \alpha F_x$$
$$Y' = -XZ + rX - Y + \alpha F_y \qquad (2)$$
$$Z' = XY - bZ + \alpha F_z$$

The strength of the forcing is controlled by the parameter α.
This external forcing could be the change in the atmospheric CO_2
concentration, for example, and a global warming associated

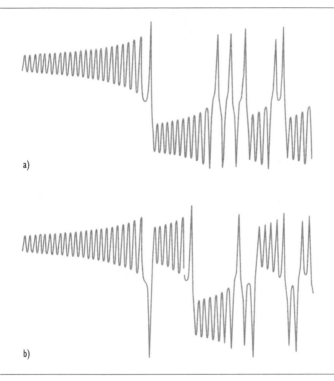

Figure 14 a) Time series of the X components of the Lorenz model from
a certain initial condition
b) The development of X over time with slightly changed
initial conditions
From Palmer (1993)

with this change. We will attempt to calculate and comprehend
how the climate of the Lorenz model changes when the parameter α varies. Although the three-component Lorenz model represents a coarse simplification of the real climate system, its results

do not depend significantly on low dimensionality. The Lorenz model is useful in so far as it clarifies the non-linearity of the climate problem.

Before we take a look at the results, it is useful to speculate how the original Lorenz attractor (Figure 13) changes when the parameter α gradually changes from zero. Erroneous linear thinking would cause us to believe that the position of the attractor would change, but the form of the attractor itself remains unchanged. The variability around the new climate state would thus not change; only a translation of the attractor would take place. However, the actual result is shown in Figure 15, in the form of the probability distribution (PDF: *probability density function*). This gives the probability for finding the state vector at any point in the X-Y plane phase space. Let us look first at the case without a forcing ($\alpha = 0$): We clearly recognize the two regimes (Figure 15 a). According to the PDF, we find the state vector in two preferred areas of the phase space, which belong to the two centers of the butterfly wings (Figure 13). Moreover, the distribution is symmetrical, i.e., the probability of finding the state vector in one of the two regimes is just as high as that of finding it in the other regime.

Figure 15b shows the results for the case with a forcing, i.e. $\alpha > 0$. In the special calculation depicted the forcing points from the one regime to the other in the X-Y plane. However, the PDF is no longer symmetrical, and the state vector is to be found more frequently in the regime to which the forcing points. Nevertheless, the coordinates of the phase space of the probability maximum are practically identical to those in the model without a forcing. In the case of minor stimuli, even the variability thus can be subject to major changes. The increasing westerlies in the winter during past decades might possibly be a sign of this non-linear behavior. As a consequence of global warming, the statistics on

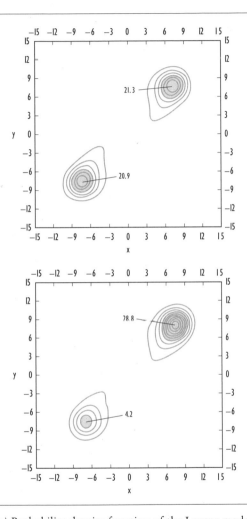

Figure 15 a) Probability density function of the Lorenz model in the
            X–Y plane without stimulus ($\alpha = 0$)
            b) Probability density function of the Lorenz model in the
            X–Y plane with a constant stimulus ($F_x = F_y = 1/\sqrt{2}$, $F_z = 0$)
            and $\alpha = 2\sqrt{2}$

            A low pass filter was applied to smooth the results. From Palmer (1993) the occurrence of certain weather situations (regimes) thus can change, which, in turn, changes the mean climate when data are averaged over longer periods. These simple considerations from the Lorenz model show that weather and climate must be considered together if the object of interest is the climate system's reaction to increased concentrations of greenhouse gases, and that the non-linearity of the atmosphere largely determines this reaction.

4.5 What Is the Gulf Stream and How Will It Behave in the Future?

The problem of the Gulf Stream is a perennial media favorite. Yet the reporting in the media does not always reflect the current state of knowledge in research. A favorite take is to pose the scenario of a new ice age should the Gulf Stream collapse. Scenarios of this kind are devoid of any scientific foundation. The possibility of regional cooling due to a weakening Gulf Stream must be set against the warming effect caused by the anthropogenic greenhouse effect. According to today's state of knowledge, the warming effect will dominate, even if the Gulf Stream weakens considerably.

What is traditionally called the Gulf Stream is the strong horizontal current on the ocean surface discovered by Ponce de Leon in 1513 (used just two decades later as the foundation of Spanish shipping route guidance), which runs along the U.S. American east coast of Florida up to North Carolina. At Cape Hatteras, 35° north, the Gulf Stream breaks away from the coast and, with a width of approximately 50 km, surges into the open Atlantic as a bundled stream making 5 km/h. After it leaves the coast the current starts to meander, assuming a serpentine path. At irregular intervals, closed rings break loose, which then circulate like lows in the atmosphere, only that they are significantly smaller and can last for many months. After around 1500 km the current loses its character as a bundled stream. The branches are then called Gulf Stream extensions; the branch reaching to Norway is also known as the North Atlantic current.

As familiar from any weather map, the Gulf Stream, too, runs parallel to lines of equal pressure: in this case the lines show the topography of the ocean surface, which can be recorded from satellites to an accuracy of just a few centimeters. A mathematically precise explanation of the Gulf Stream was not provided until 1947, by Sverdrup. The causes are the rotating movements forced by atmospheric circulation (trade winds from eastern directions in the tropics, and from westerly winds in the mid-latitudes) and the effect of the Earth's rotation on the movement of water (Coriolis force), which increases as water moves north. The Gulf Stream and its extensions guide warm, tropical water toward Europe and are thus responsible for the unusually mild climate in western and northern Europe compared to other areas of the same geographic latitude (like Canada, for instance).

In recent years another, vertical aspect of the Atlantic ocean circulation has attracted public interest, known as the thermohaline circulation (see section 1.2). The warm water on the surface of the North Atlantic, the lion's share of which flows to the north, after cooling down in the winter and sinking to the deep layers as a consequence of convection, returns to the southern ocean. This vertical circulation is popularly referred to as the 'oceanic conveyor' or 'meridional overturning circulation.' The heat transport toward the north associated with this phenomenon is estimated at about one billion megawatts, amounting to 300 million kilowatt hours per second. A collapse of the thermohaline circulation would cause temperatures in central Europe to fall by an average of around 1–2°C yearly, whereby the cooling effect would be noticeable primarily in the winter. Our mild climate thus apparently can be traced back not only to the thermohaline circulation, but also to considerable heat transport by the atmosphere. Furthermore, its wind-driven component would keep the Gulf Stream from grinding to a complete halt even if the thermohaline component were to collapse.

To keep the thermohaline circulation in motion, sufficient salty water is required at high latitudes, which can become very cold due to the drop in the freezing point. In today's climate we can regard this mechanism as a self-sustaining pump: the more deep water is cooled, the more salt-rich water from the subtropics is pulled toward the north on the surface. With various computer models it has been shown that a large influx of fresh water, like when the continental ice sheets melted after the ice age, for instance, would be enough to disrupt the thermohaline circulation. For the future development of the climate, an increase in the transport of water vapor from the tropics to the pole as a consequence of global warming, or a change in the mass balance on Greenland, could produce such a change in propulsion.

The thermohaline circulation is a very sensitive current system, the stability of which was examined theoretically by means of a conceptual model by Stommel in 1961. Stommel's non-linear model consists of two boxes, an equatorial box and a polar box. The rate of flow that describes the strength of the thermohaline circulation is directed from high to low pressure and driven by differences in density between the boxes, which are imprinted on the system through the processes of interaction with the atmosphere. In dimensionless form (with the primed variables designating the temporal derivatives), the equations read:

$$T' = \eta_1 - T(1 + |\psi|) \qquad (3)$$
$$S' = \eta_2 - S(\eta_3 + |\psi|)$$

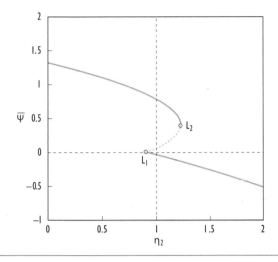

Figure 16 A bifurcation diagram showing the stationary values of the rate of flow ψ of the model (3) for various values of the parameter η_2 and fixed $\eta_1 = 3.0$ and $\eta_3 = 0.3$, respectively

The dotted branch of the solution is instable. From Dijkstra (2000)

In these equations T and S are the dimensionless differences in temperature and salinity between the two boxes, and $\psi=T-S$ the dimensionless rate of flow. η_1, η_2 and η_3 are dimensionless parameters. The parameter η_2 is of particular importance here, as it describes the changes in the influx of fresh water on the surface, which may be subject to considerable influence by the anthropogenic greenhouse effect. In particular, the possibility of Greenland's ice melting would be covered by this parameter. Now let us observe the behavior of the system when the parameter η_2 is varied (Figure 16).

For values of the parameter η_2 up to point L_1, the stable solution is called the TH solution, which is characterized by descent in the polar box and upwelling in the equatorial box. This state corresponds to today's climate with a stable thermohaline circulation and a relatively strong heat transport directed to the north. For values of the parameter η_2 greater than L_2, the stable solution is known as the SA solution, which is marked by descent at the equator. This solution would correspond to an inversion of the thermohaline circulation, and thus would have far-reaching consequences for the climate over the North Atlantic and Europe. Between the two values L_1 and L_2 there are two states of equilibrium, i.e., the system can take on two different states depending on the initial conditions. The assumption is that at the end of the last ice age the thermohaline circulation was located in the vicinity of this bipolar regime, through which strong reversals occurred repeatedly, also reflected in the temperatures in Europe. But where are we today? This question is a matter of intense discussion in science today. Should today's thermohaline circulation be located near a bifurcation point, i.e., a tipping point, even relatively minor disturbances could be enough to disrupt it.

The complex climate models calculated today may provide different answers when they are calculated with increased concentrations of greenhouse gases in the atmosphere, but most models simulate an only moderate weakening of the thermohaline circulation for the year 2100 (~25% on average, see IPCC 2007), which results in a delay of the warming over the North Atlantic. The models thus agree that the system is still relatively far from a bifurcation. None of the models simulate a cooling over Europe by 2100. Greenhouse warming develops more quickly and exceeds the cooling tendency through the weakening Gulf Stream.

In so far ice age scenarios lack any foundation at all. Even if the thermohaline circulation were to collapse today, only moderate cooling in central Europe would result, as discussed above. Results of simplified models have been presented that show a complete stop in the thermohaline circulation if the concentration of greenhouse gases in the atmosphere increases rapidly. Therefore the question as to the future of the Gulf Stream system is still a matter of very controversial scientific discussion. Yet a complete collapse of the thermohaline circulation within the next decades accompanied by a cooling of Europe appears exceedingly improbable.

An attempt to reconstruct the strength of the thermohaline circulation is possible on the basis of the observations of ocean surface temperatures collected over the past hundred years. From this it is then possible to derive the extent to which a weakening of the thermohaline circulation already can be observed today. The idea goes something like this: the thermohaline circulation is associated with a northward heat transport in the Atlantic. Through this the North Atlantic is warmed and the South Atlantic cooled. Changes in the strength of the thermohaline circulation should thus be measurable in the difference between the temperatures at the surface of the North Atlantic versus those of the South Atlantic. A strong thermohaline circulation is connected with a major temperature differential; a weak circulation with a minor one. In so far this temperature opposition between the North and the South Atlantic serves as a kind of fingerprint for the strength of the thermohaline circulation. The advantage of this fingerprint is that observations of ocean surface temperatures are available for relatively long periods of time, so that it would be possible to recognize a long-term trend. Moreover, no elaborate current measurements are necessary if the temperature fingerprint actually measures the strength of the thermohaline circulation.

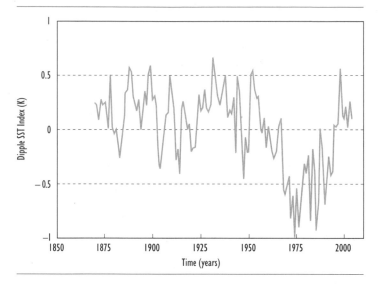

Figure 17 The temperature contrast between the North and South
Atlantic, which can be regarded as a measure of the strength
of the thermohaline circulation

Marked fluctuations can be detected, especially on the time scale of decades, but
no long-term weakening of the thermohaline circulation, especially not in recent
decades

Figure 17 shows how this temperature differential has developed over time. Marked, long period fluctuations can be detected,
but no long persistent trend that would suggest a weakening of
the thermohaline circulation. A strengthening of the temperature differential actually can be observed in recent decades. This
result shows that a clear weakening of the thermohaline circulation can not yet be measured. This result is also consistent with
the climate model simulations on global change, which do not
allow any significant change to be expected today.

The Climate of the 20th and 21st Centuries

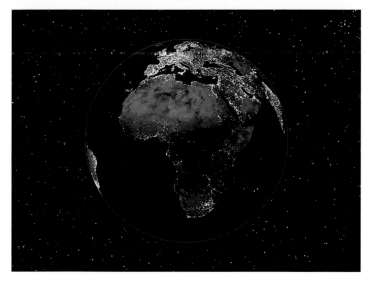

Figure 1   The Earth by night as seen from outer space. The lights in Europe can be clearly recognized, while most of Africa remains dark. This difference illustrates the enormous consumption of energy in industrial nations as compared to the developing world.

Figure 2 All over the world, energy is generated primarily by burning fossil fuels (petroleum, natural gas, coal). This releases large amounts of carbon dioxide into the atmosphere, heating up the Earth. Smoking chimneys are a symbol of energy consumption and thus for man's influence on the climate.

Figure 3 Arctic sea ice is receding ever further. A comparison of the ice cover in 1979 (above) with that in 2005 (below) makes this clear. In particular the area known as the Northeast Passage has become free of ice.

Figure 4 Glaciers are receding all over the world. The example of the Pasterze Glacier in Austria offers a particularly good illustration of this phenomenon. While a distinctive glacier tongue could be recognized around 100 years ago (above), this has disappeared completely today (below).

5 Man's Influence on the Climate

5.1 The Intergovernmental Panel on Climate Change (IPCC)

The question as to whether man is changing the climate has long been answered by international climate research. Today there is practically nobody left who would contest the existence of climate change. Climate change is thus in full swing, and its signs are unmistakable (see Figure 18). A decisive role in the evaluation and communication of the scientific results falls to the Intergovernmental Panel on Climate Change (IPCC). The Intergovernmental Panel for Climate Change was founded in 1988 by the World Meteorological Organization (WMO) and the United Nations Environment Programme (UNEP). Its brief is, first, to describe the state of scientific knowledge with regard to global climate change, and second, to advise international policymakers. The Intergovernmental Panel on Climate Change and Albert Arnold (Al) Gore Jr. were awarded the Nobel Peace Prize in 2007 'for their efforts to build up and disseminate greater knowledge about man-made climate change, and to lay the foundations for the measures that are needed to counteract such change.'

Since 1990 the IPCC has provided a number of assessment reports, which have become standard references used frequently by political decision makers, scientists and other experts (the reports are available in the Internet at www.ipcc.ch). Hundreds of the world's leading climate scientists have contributed to the

IPCC reports. Even the report back in 1995 stated that 'The balance of evidence suggests a discernible human influence on global climate.' The report of 2001 states, 'An increasing body of observations gives a collective picture of a warming world and other changes in the climate system.' Finally, the last (fourth) report published in 2007 states, 'The understanding of anthropogenic warming and cooling influences on climate has improved… leading to very high confidence that the global average net effect of human activities since 1750 has been one of warming.' Apparently there is a major consensus in international climate research that our climate is changing and that man is partially responsible for this. The next section will describe a number of important observations that contributed to this consensus.

5.2 What Changes Can Already Be Seen Today?

As described above, since the beginning of industrialization we have observed a strong increase in the concentrations of greenhouse gases in the atmosphere, above all of CO_2. This increase reinforces the greenhouse effect and results in global warming on the Earth's surface and in the lower layers of its atmosphere. The question thus arises as to what climate changes already can be observed today. In posing this question we must not lose sight of the fact that the climate, as a consequence of its inertia, always reacts to external stimuli with a delay of several decades. Thus we cannot presume to observe the climate system's entire reaction to our behavior today. Yet strong warming of the Earth is already evident, both globally and on the Northern Hemisphere (see Figure 18). Roughly speaking, the increase in global temperature over the past hundred years amounts to approximately 0.8°C, around 0.6°C of which were added just in the last 30 years.

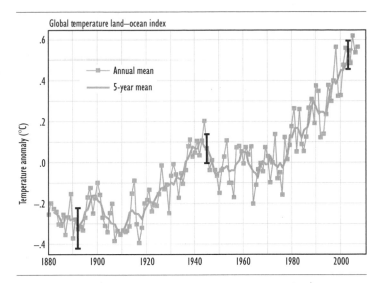

Figure 18 Global mean land-ocean temperature index, 1880 to 2007
The thin green line is the annual mean and the thick green line is the five-year mean. The black bars show uncertainty estimates (redrawn after NASA, http://data.giss.nasa.gov/gistemp/graphs/)

The year 2005 was the warmest year recorded since the beginning of temperature measurements. The argument that most of the warming took place in the first half of the 20th century when greenhouse gas concentrations rose only quite weakly thus can be rejected once and for all. What is also interesting is the fact that the temperature has not risen continuously, but has shown a wide range of fluctuation. This behavior makes clear that the climate is influenced not only by humans, but by a multitude of other natural factors as well. Yet in the past decades it appears that the human influence increasingly has become the decisive factor, as seen in the rapid warming over the last decades.

Warming over the continents was considerably stronger than over the oceans. In Germany, the warming over the last hundred years amounted to approximately 1°C, confirming a characteristic land-ocean contrast: land regions clearly warm up more intensively than the ocean surface. Of course the analyses of the measurements also took into consideration that measurement methods have changed over the course of time, as well as the existence of a city effect, through which metropolitan areas warm especially intensively. The 0.8°C change in the global mean temperature in the last hundred years thus must be understood as an adjusted value, for which the known systematic errors are already corrected.

Reconstructions of the temperature of the Northern Hemisphere over the last thousand years show how extraordinary the warming of the last hundred years has been in comparison to the changes in the centuries before. Even accounting for the fact that the temperatures before 1900 were derived primarily from indirect methods (like, for instance, the analysis of ice cores, tree rings or corals), and allowing for the maximum degree of uncertainty in the temperatures thus determined, the decade between 1990 and 1999 was the warmest in the last thousand years. Combined with additional statistical and model-based analyses (fingerprinting methods), it is possible today to claim with very high probability (over 95%) that the temperature rise observed over the last decades can be traced back to humans. In the past there have been repeated climate fluctuations that were not a result of human activity, such as the Medieval Warm Period or the Little Ice Age, yet all of these were considerably weaker than the rise in temperature over the past decades, at least on the global scale.

Temperature is not the only evidence for the fact that our climate is changing, however: the ice and snow cover of the Earth has diminished as well. The snow cover of the Northern

Hemisphere has decreased by approximately 10% since 1960. Mountain glaciers have receded all over the world. The extent of sea ice on the Northern Hemisphere has also diminished and the ice has become considerably thinner. Since the start of regular satellite measurements, the summer extent of the Arctic sea ice was never as small as in the year 2007, when it fell well below 4 million km² for the first time.

Since warming in the Alps was around double the level of the global mean, several glaciers in this region have already lost around 50% of their mass during the last hundred years. Sea level increased by around 20 cm in the 20th century, in part because of the melting glaciers and in part because the heat content of the oceans increased considerably, leading to the thermal expansion of ocean water. The frequency of extreme precipitation in the mid-latitudes and higher latitudes of the Northern Hemisphere increased, as did summer droughts.

Another phenomenon also appears to be changing as a consequence of global warming: the intensity of hurricanes increased during recent decades. The preliminary stages to hurricanes are called tropical storms. We differentiate among six categories of tropical cyclones: tropical storms and five hurricane categories of ascending strengths. Category five designates the most severe. The year 2005 had the most tropical cyclones in the Atlantic region since measurements began in 1850. While the previous record of the year 1933 registered 21 tropical cyclones, 28 tropical cyclones were observed in the Atlantic Sector (Figure 19) in 2005. Particularly striking is the major increase in the number of strong hurricanes (categories 3 – 5) in recent decades. However, as natural fluctuations are quite great from one decade to the next, the current intensification of tropical cyclones can be seen as no more than an indication of the anthropogenic influence.

Tropical cyclones arise only over areas in which the ocean

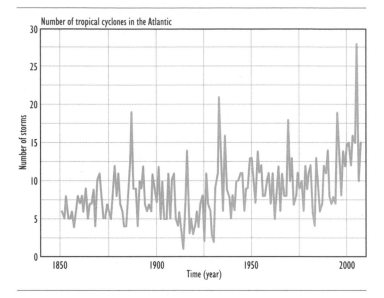

Figure 19 Number of tropical cyclones 1850–2007
Data from NOAA

temperature is above 26.5°C. Other criteria must also be fulfilled, like, for instance, a deep mixed layer in the ocean, and minimal vertical wind shear. The accumulation and even intensification of tropical cyclones in the Atlantic sector is accompanied by a rise in the temperature of the ocean surface of the tropical Atlantic, the region over which hurricanes develop. The rise in ocean temperature in recent years has two causes. First, there is a natural cycle with a period of several decades (60 – 80 years), which can be traced back to fluctuations in the circulation of the Gulf Stream (thermohaline circulation). At present we find ourselves in the warm phase of this cycle. Second, global warming exerts an additional influence on the temperature of the ocean, such that by now the ocean temperature of the tropical Atlantic has reached the highest values ever measured. In summer 2005, for instance, the temperature of the ocean surface was up to 2°C higher than the long-term mean value. In so far it is plausible, but not yet proven, that the long-term warming of the Earth is at least partially responsible for the intensification of tropical cyclones over the Atlantic observed in recent decades.

Yet there are also areas, especially on the Southern Hemisphere, that exhibit no noteworthy changes. Climate models show that this is to be expected, too, because of the lower share of land, and because the Antarctic Ocean is warmed relatively little at the surface as a consequence of strong vertical mixing. Yet overall the signs of anthropogenic climate change are unmistakable, as can be read in detail in the reports by the IPCC.

5.3 Whose Fault Are the Changes?

Over and again the question arises as to the role of natural factors like the sun in the warming of the Earth. The possible role of fluctuations in solar radiation on the climate has long been a topic of discussion in climate research. Solar radiation is subject to fluctuations on the time scale of decades, many of which are associated with the activity of sunspots. Thus a high number of sunspots means an increase in solar radiation, and is also connected with a slight shift in the solar spectrum toward the short wave (UV) range. There are two known cycles. The first is known as the Schwalbe cycle, which has a period of eleven years and a directly measured amplitude of approximately 0.1%, and the second is called the Gleissberg cycle, with a period of about 80 years and an estimated amplitude of approximately 0.2 – 0.3% of total irradiation. However, the net radiative forcing by the sun since 1750 amounts to only about 0.1 W/m². By comparison: the additional greenhouse effect caused by increased concentrations of carbon dioxide, methane, chlorofluorocarbons and nitrous oxide currently amounts to approximately 2.6 W/m². Furthermore, the majority of the rise in the solar constant took place in the first half of the 20th century; in fact, it has been dropping gradually since 1940.

An influence of cosmic radiation on the climate is also frequently postulated. Ionization is believed to facilitate the formation of aerosols, which thus stimulates cloud formation. The changed cloud cover, in turn, is believed to influence the temperature of the Earth (see section 7.2). Yet to date there is no scientific evidence supporting this hypothesis using reproducible statistical methods or calculations with complex climate models. The influence of cosmic radiation is thus not considered in any of the climate models recognized worldwide.

How strongly do solar fluctuations and other natural factors, such as a change in volcanic activity, influence the climate? And how strong is humans' influence on the climate? In order to answer these questions, two different simulations were carried out using a number of complex climate models, which are presented in Figure 20. In the first simulation only natural factors

Figure 20 a) Global mean surface temperature anomalies relative to the period 1901 to 1950, as observed (black line) and as obtained from simulations with both anthropogenic and natural forcings

The thick green curve shows the multi-model ensemble mean and the green shading shows the range of the individual simulations. Vertical green lines indicate the timing of major volcanic events

b) As in a), except that the simulated global mean temperature anomalies are for natural forcings only
(redrawn after IPCC 2007)

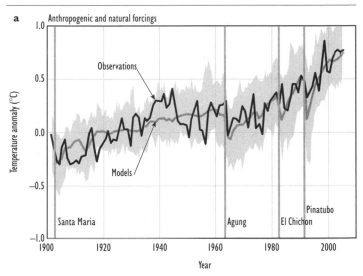

a Anthropogenic and natural forcings

Observations

Models

Santa Maria   Agung   El Chichon   Pinatubo

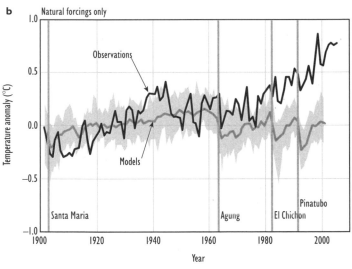

b Natural forcings only

Observations

Models

Santa Maria   Agung   El Chichon   Pinatubo

were considered, in the second both natural and anthropogenic factors. The two simulations were computed in ensemble mode, meaning that they were carried out with varying initial conditions. This allows us to estimate the uncertainty in the model results due to the internal variability of the climate. The range of the individual simulations is represented as shading, the ensemble mean as a thick line. The latter is an estimate of the externally forced climate variability.

The results of the models show that natural factors, especially the increase in the intensity of the sun, certainly can explain part of the observed warming; however, this accounts for around 0.2°C, or just about one quarter. The variability of the sun alone thus cannot be responsible for the rise in temperatures of approximately 0.8°C observed over the last hundred years. In particular, natural factors can explain only a certain rise in temperature in the first half of the 20th century, but not the strong warming trend in recent decades. This was also to be expected, as the rise in solar radiation was largely over by 1940. Only the simulation in which both the natural and the anthropogenic factors were accounted for produces satisfactory correspondence with the observations. The predominant share of the Earth's warming, especially in the last decades, is thus obviously caused by man.

Moreover, according to climate model simulations of the anthropogenic greenhouse effect combined with the influence of the anthropogenic 'sulfate aerosols' – these also originate primarily from burning fossil fuels, but exert a cooling effect – the human impacts create a characteristic horizontal spatial warming pattern (fingerprint), which is recognizable in the observations and clearly different from the pattern of natural (for instance, solar) forcing. The same is true for the temporal evolution of the warming over time, as shown above. Finally, the vertical structure of the observed temperature change cannot be explained by solar forcing. While the troposphere warmed, the stratosphere cooled during the recent decades. This is the characteristic pattern expected from the anthropogenic greenhouse effect. Solar forcing would not produce cooling of the stratosphere. Thus the entire time-space structure of the observed warming points to man as the main cause of the Earth's warming in the 20th century.

5.4 Mankind's Fingerprint

Because of the great inertia of the climate system it is important to recognize man's influence on the climate as soon as possible, so that countermeasures can be introduced in time. The first proof was submitted in the year 1995. As described above and expressed simplistically, CO_2 has the effect of glass in a greenhouse: it lets solar radiation through, but prevents part of the Earth's heat from being radiated into space, so that an increasing concentration of CO_2 entails the danger of global warming. Additional factors influencing the climate include sulfate aerosols and soot. Sulfate aerosol particles are created by industrial sulfate emissions into the atmosphere, but also in the combustion of fossil fuels. They reflect part of the solar irradiation and thus cause regional cooling. By means of simulations using coupled ocean-atmosphere-sea ice models and improved signal recognition methods, it was in fact possible to demonstrate the high probability that humans are responsible for the warming of the Earth observed in the last decades. The proof of anthropogenic climate change is an important step toward motivating international politics to institute climate protection measures.

Time-dependent climate change simulations (see chapter 6) and long 'control' simulations without external forcing serve as valuable tools supplying statistical proof for the anthropogenic greenhouse effect. They provide a prediction of the spatial and temporal structure of anthropogenic climate change, and allow the statistical properties of natural climate variability – 'climate noise' – to be estimated. The question this raises is the extent to which the observed space-time structure of the warming over the last decades can be reconciled with the patterns of natural climate variability. If the observed warming lies outside of a defined confidence interval, in which, for example, 95% of natural climate fluctuations take place, there is only a 5% statistical chance that this warming represents natural climate fluctuation. Inversely, there is a probability of 95% that this warming can be traced back to mankind.

One could use only the mean global temperature as statistical proof of the anthropogenic greenhouse effect, as in our interpretation above on the basis of Figures 18 and 20. However, an anthropogenic greenhouse gas signal can be separated from climate noise – that is, from natural fluctuations – sooner and more reliably if the spatial structure of anthropogenic climate change, its 'fingerprint,' is considered and determined by means of model simulations. The observed temperature change is then compared with this fingerprint so that the problem can be reduced to a scalar detection variable, i.e., a single number. The chances of detecting a climate change improve even further when the fingerprint is changed such that it suppresses natural climate fluctuations ('optimal' fingerprint). This method weights the components of climate change that exhibit strong noise more weakly than those which are associated with lower natural climate fluctuations.

The fingerprint method was applied to the spatial pattern of linear trends (increases in the slopes adjusted to the local time series) of the air temperature near the surface over the course of

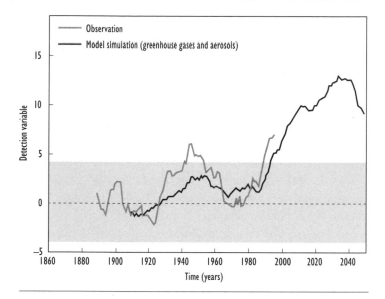

Figure 21 Development of the detection variable in 30-year temperature trends in observations (green) and in the model world (black)

The detection variable is marked against the last year of the trends (thus '1994' designates the detection variable for the period 1965–1994). The gray area designates the 95% confidence interval for those fluctuations in the detection variable that are due to the natural variability of the climate

thirty years. With this the spatial component is complemented by a temporal component, as trends are observed over time. Long-term, homogenized temperature measurements describe the observed climate evolution. Analysis is performed only for areas where reliable temperature trends can be calculated. Here the control experiment of a model calculated without external forcing is used to optimize the fingerprint. Additional control simulations using other models serve to assess those fluctuations in detection variables due to climate noise even more accurately. Figure 21 shows that the detection variable recently has begun to deviate from natural noise.

Therefore we can reject the hypothesis that the rise in the near-surface air temperature observed in the years 1965 – 1994 was part of natural variability with a risk of less than 5% (formally, actually only 2.5%, because a 'univariate' test was performed toward positive swings in the detection variable). If the last years are included in our consideration, the probability of error drops even further, as the planet has warmed further and this additional warming resembles the fingerprint. The black curve in Figure 21 shows the course of the detection variable as forecast by the climate model (in this case the fluctuations are reduced by averaging two experiments using different initial conditions). The model's forecast is consistent with the anomalous level of the detection variable in the observations toward the end of the last century.

Nevertheless this result depends on the assumption that the estimation of natural variability by the model and observation data is sufficiently accurate. The significant temperature trend in the first half of the 20th century around 1945 presumably can be explained in part by the anthropogenic greenhouse gas signal already developing at that time, but to a larger degree by an extreme event of natural variability, presumably increased solar irradiation. Paleoclimate data suggest that this may have been the strongest thirty-year trend in the Northern Hemisphere during the last 500 years. Yet the fact that the estimation of variability turned out to be different for the various climate models shows that the estimation of natural climate variability is still subject to some uncertainties.

Proving that climate change is statistically significant, however, is not sufficient to establish a causal connection between this change and changes in the concentration of greenhouse gases. In order to link climate change with the change in the concentration of greenhouse gases, we must be able to exclude all other causes of an externally induced climate variation. In spite of the high uncertainty in our knowledge of the history of solar fluctuations, volcanic eruptions and other events, complete with their effects on the climate, an explanation that attributes the current warming trend to these factors appears extremely improbable. The strong consistency of observations with the results of the model, not to mention the results of other calculations which investigate such factors as the vertical structure of the temperature distribution in the atmosphere, suggest that the significant warming of the Earth observed, especially in the last decades, is indeed connected with the rise in concentrations of greenhouse gases.

6 Climate Change Scenarios for the Future

6.1 The Inertia of the Climate

Inertia is an inherent property of the climate. Several effects of anthropogenic climate change therefore manifest themselves only gradually. Some of these effects could be irreversible once certain threshold values are exceeded. Yet the location of these thresholds is known only imprecisely. Because of CO_2's long lifetime (about 100 years), stabilizing CO_2 emissions at today's level would not result in a stable CO_2 concentration, even though stabilizing some of the shorter-lived greenhouse gases (like methane) does lead to a stabilization of their concentrations. Stabilizing the concentration of CO_2 at a certain level would mean reducing the global net emissions of CO_2 to a fraction of their current quantities. The lower the level of stabilization to which we aspire, the earlier we must begin reducing global net emissions.

The inertia of the climate means that the climate will continue to change long after the concentration of CO_2 has stabilized. Climate models show that the surface air temperature will continue to rise by several tenths of a degree for at least a century after the atmospheric CO_2 concentration has been stabilized, and sea levels by many decimeters over centuries. The slow transport of heat to the deep ocean and the slow reaction of the ice sheets mean that it will take over a millennium until a new state of equilibrium for sea level is achieved (Figure 22).

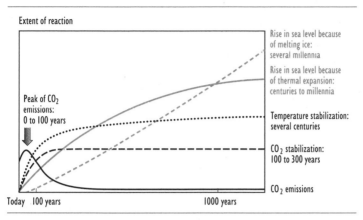

Extent of reaction

Peak of CO₂
emissions:
0 to 100 years

Rise in sea level because
of melting ice:
several millennia

Rise in sea level because
of thermal expansion:
centuries to millennia

Temperature stabilization:
several centuries

CO₂ stabilization:
100 to 300 years

CO₂ emissions

Today   100 years   1000 years

Figure 22 Even after the reduction of CO₂ emissions and the stabilization of the CO₂ concentration in the atmosphere, surface
temperatures continue to rise by a few tenths of a degree per
century for a century or more

Thermal expansion of the oceans continues long after CO₂ emissions are
reduced, and melting ice sheets continue to contribute to rising sea levels for
many centuries. This figure is a generic illustration of a stabilization at a given
level between 450 and 1000 ppm and thus does not bear any concrete numbers
on the axis of effects (from IPCC 2001b)

Several changes in the climate system that are plausible in predictions beyond the 21st century would be potentially irreversible. Fundamental changes in the patterns of ocean circulation,
like the Gulf Stream, for instance, caused by major melting of
the Greenland ice sheet, could not be reversed over a period of
many human generations. The threshold value for fundamental changes in ocean circulation can be reached through only
minor warming, if it takes place rapidly rather than gradually.
Inertia and the potential for irreversibility are important reasons
why foresighted measures of adaptation and mitigation are
imperative.

6.2 What Will the Future Bring?

Through the combustion of fossil fuels and other human activities the concentrations of atmospheric greenhouse gases, for example like carbon dioxide and methane increase, but so does the share of the particles (e.g. sulfate aerosols) that reflect a portion of sunlight and thus work to counter the anthropogenic greenhouse effect. The future consequences for the Earth's climate can be estimated using computer simulations. As discussed above, global climate models are programmed to do this by quantitatively describing the interaction between physical processes in the atmosphere, ocean, sea ice, and land surfaces. The input parameters required by any given model include the concentrations of the most important long lasting greenhouse gases, while the concentrations of short-lived aerosols, which are closely linked with internal processes like the formation of clouds and precipitation, are generally calculated from the emissions within the climate model.

Yet the results of the climate models are crucially dependent on the given 'scenario,' i.e., on the assumptions about the future development of such variables as the world population, industrialization, and the consumption of fossil fuels. (Figure 23 shows the scenarios for CO_2 concentration, an input used to drive the climate models). However, different climate models yield differing warming forecasts, even when they are calculated using the same scenario. For the time period 1990 – 2100, the IPCC gives a bandwidth of 1.1 – 6.4°C for the global temperature trend (Figure 24). This relatively wide range results first from the uncertainty about the future emissions of trace gases, and second from the model uncertainty. Nevertheless, even with a very strong reduction in trace gas emissions by the year 2100, we must figure on additional warming because of the inertia of the climate. Even

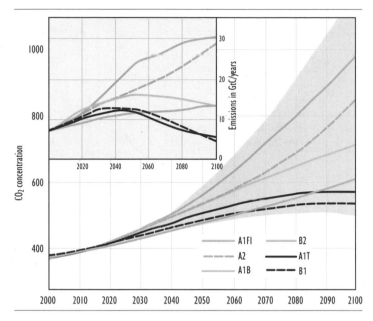

Figure 23 CO₂ emission (small image) and concentration scenarios
(SRES) used in climate model projections of the future climate
(redrawn after www.hamburger-bildungsserver.de)

in the impossible event that the greenhouse gas concentrations
can be stabilized at year 2000 levels, the Earth will warm by an
additional 0.3–0.9°C. However, we can still prevent more intense
warming.

In the extreme case the Earth would approach an average
mean temperature of approximately 20°C in the year 2100. As
Figure 25 makes clear, this temperature would lie considerably
higher than is typical for historical interglacial periods. The last
major interglacial period, the Eemian interglacial, which took
place approximately 125 000 years ago, was considerably colder.

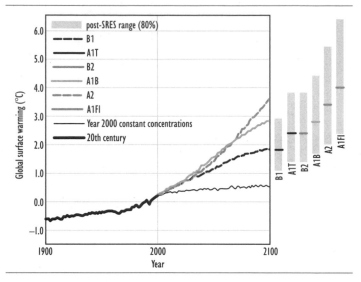

Figure 24 Projected changes in global mean surface temperature during
the 21st century

Solid lines are multi-model global averages of surface warming for scenarios A2,
A1B and B1, shown as continuations of the 20th-century simulations. The lower
line is not a scenario, but stands for simulations where atmospheric
concentrations are held constant at year 2000 values. The bars at the right of the
figure indicate the best estimate (solid line within each bar) and the likely range
assessed for the six SRES 'marker scenarios' at 2090–2099. All temperatures are
relative to the period 1980–1999 (redrawn after IPCC 2007)

According to what we know today, a global mean temperature
of approximately 20°C has not been experienced for at least one
million years. Moreover, warming of around 5°C would correspond to the temperature difference between the last ice age
20 000 years ago and today. However, in contrast to this the forecasted greenhouse warming would develop over just one hundred
years. A global climate change of this speed has never taken place
in the history of mankind.

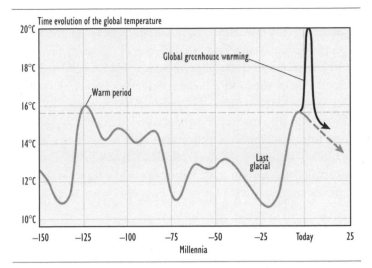

Figure 25 The maximal possible temperature change by 2100 compared to the temperatures of the last 150 000 years

Climate models can never provide more than an approximate description of the very complex real climate system. The validity of the model generally decreases directly in keeping with the scale of the area under consideration. For example, regional details within Germany can be comprehended with less precision than can differences between northern and southern Europe. The main reason is that grid of existing global climate models is still relatively coarse, with a few hundred kilometers between grid points, thus precluding, for instance, high-resolution portrayals of mountains like the Alps, or detailed representation of small-scale processes like the formation of clouds and precipitation. This is compounded by the fact that today's models are still incomplete. Factors like possible changes in vegetation in the climate of the future generally receive just as little consideration as changes in the mass of the continental ice sheets. For instance, vegetation could change as a consequence of the climate changes sketched above, and this change, in turn, could have effects on the temperature of the land's surface. Such feedback mechanisms from vegetation dynamics will presumably receive more consideration in the next generation of climate models, as will the interaction with chemical processes in the atmosphere. Yet for the record, in spite of the uncertainties described above, models are capable of calculating the large-scale and long-term changes in the climate relatively reliably, as shown in Figure 20 above.

6.3   The Climate in Europe in the Mid-21st Century

A model developed by the Max Planck Institute for Meteorology will now be analyzed with a view to the changes expected over Europe. The results are characteristic of many climate models (see the last IPCC report published in 2007). Global warming entails an increase of water vapor in the atmosphere as well as increased transport of water vapor from the oceans to the continents, and thus an increase in precipitation over land. Yet the changes in precipitation differ widely from one region to the next. Generally, more precipitation falls in high latitudes and in parts of the tropics, while the drier subtropics experience even less rain. Thus the differences increase between the moist and dry climates on Earth. This statement is also valid for the climate in Europe. However, the tendencies for precipitation are very different in the winter months than in the summer. While summer precipitation is decreasing nearly everywhere in Europe, in winter a marked north-south differential is predicted, with a decrease in the lower-precipitation areas of southern Europe and an increase in rainier central and northern Europe. This increase is associated with intensified winter storm activity over the northeast Atlantic and stronger westerlies, bringing moist air from the Atlantic. Striking is the increase in strong precipitation events and the increased probability of attendant flooding. This is true in part even for the Mediterranean region, where the average amount of precipitation is dropping. The presumptive cause is the higher water vapor content of the atmosphere, which makes greater amounts of precipitation possible during extreme weather events than does today's climate.

Regional and seasonal differences can also be detected in the simulated temperature changes. While the greatest warming in summer, of up to 2.5°C in Spain, is simulated by the middle of this century, the warming trends in winter are particularly strong (up to 5°C) in regions like northern Scandinavia and Russia, where less snow is expected to fall as a consequence of warmer temperatures. These tendencies also can be recognized to a lesser extent in annual mean temperatures. Over all, in the mid-latitudes there will typically be approximately 20 fewer days below freezing by the middle of this century; the number of hot days per year, with highs of more than 30°C, will also increase by approximately 20.

One of the important questions with regard to the climate shift is whether flooding will become more frequent in Europe. Various causes work together to increase or decrease river flooding. The first cause is interventions in the flow rate of rivers and their feeders: river regulation, the construction of embankments and irrigation systems, and changes in the land use of the river catchment (for example deforestation). These factors differ from river to river and their impact for the future is difficult to assess.

A second crucial influence on the frequency and severity of flooding is climate change, and particularly a rise in the frequency of extreme precipitation events. All models predict that the average warming caused by an increase in greenhouse gases will be accompanied by an intensified water cycle and thus a higher global average precipitation level. What this could mean for individual regions is illustrated on the basis of two calculations using a global model of the atmosphere. First today's climate was simulated for the period 1970 – 1999, by using the observed concentrations of greenhouse gases. Then the future climate was simulated for the period 2060 – 2089, whereby estimations for future concentrations of greenhouse gases were extrapolated in keeping with Scenario IS92a of the IPCC. The simulation carried out here is known as a 'time slice experiment,' in which the conditions simulated with a coarse resolution model (horizontal resolution approximately 300 km) are fed into a relatively high-resolution model (horizontal resolution approximately 100 km). The coarse resolution model is simulated continuously for the period 1860 – 2100, as described in the previous section. The results for ocean surface temperatures and sea ice cover in the periods ('time slices') of interest (1970 – 1999 and 2060 – 2089) are then fed into the high-resolution model. This method allows higher regional precision to be achieved without calculating the costly high-resolution model for the entire period.

In northern and central Europe the number of days with precipitation over 20 mm/day is increasing considerably – in other words, extremely strong precipitation events are becoming more frequent. A comparison with the simulation of today's climate shows that in certain areas the number of days with strong precipitation has even doubled (on the Norwegian coast, for example). This tendency is already apparent at the Hohenpeissenberg station in the Alpine foothills, where the number of strong precipitation events has more or less doubled in the last hundred years.

In order to study the influence of the changed precipitation on the rivers, the results presented above were input into a model of lateral runoff for land surfaces, whereby interventions by man – through the construction of embankments, or river regulation, for example – were disregarded in order to determine the pure climate effect. In all regions where both average precipitation and extreme precipitation events are increasing, strong flooding has become more frequent. This is especially true for northern Europe and parts of central Europe.

Moreover, the model also simulates lengthier summer droughts for large parts of Europe. The summer extremes will thus increase in both directions: longer droughts and more strong rainfalls. These assessments from model calculations can only offer an indication of what the future development might be. It is hoped that higher resolution models will produce even more accurate calculations. Calculations using global climate models with typical grid widths of 10 km should be possible in the next two years. Regional climate models of correspondingly high resolution, like the ones calculated for Europe, confirm global model simulations to the extent that we do indeed expect longer dry periods in the summer, but also more extreme rainfalls.

6.4 How Much Will the Sea Level Rise?

The mean sea level is subject to fluctuations on various time scales. In addition to short-term oscillations through wind or tides, long-term changes occur due to geological and climatological processes. The latter can then be divided into climate changes caused by nature and those caused by man. This section describes the individual factors that exert a significant influence on long-term changes in the sea level.

First, the warming of ocean water through the anthropogenic greenhouse effect causes an expansion of the water column, thus resulting in a rise in the sea level. A warming of the entire water column by 1°C, for example, would cause the sea level to rise by about 50 cm. Such an evenly spread warming of the entire water column within a short period of time is unrealistic, however. Yet because the deep ocean warms much more slowly than the ocean surface, the already extremely slow vertical exchange between these layers declines, with the overall effect of slowing the rise in sea levels. The value stated is therefore a mere estimate of the how much 'thermal expansion' contributes to a rise in sea level.

Second, the large ice sheets can change. The Antarctic represents the world's largest warehouse of freshwater outside the oceans. Its volume is estimated at about 25 million km^3; if it melted completely, sea level would rise by around 60 m. The second largest warehouse is the Greenland ice sheet. Its volume is estimated at just less than 3 million km^3; its melting would mean an increase of around 7 m in sea level. No unambiguous observations have been submitted that would suggest a long-term melting or growth of the ice sheets in the Antarctic or Greenland, but satellite measurements from recent years show truly alarming rates of melting in Greenland. Moreover it is not at all clear how the stability of the large ice sheets might be affected by changes in the albedo of the ice sheets, i.e., their capacity to reflect sunlight, caused by such pollutants as soot. As a consequence of summer thawing and increasing infiltration by meltwater, the large ice sheets could react more quickly than was assumed up to this time.

Third, for about 100 years a contraction of the mountain glaciers has been observed. This melting leads directly to an increase in sea level, as the meltwaters ultimately end up in the ocean. The volume of the glaciers amounts to about 0.1 million km³. If all of the mountain glaciers melted completely, sea level would rise by up to 0.4 m. Incidentally, the melting of the sea ice and shelf ice already floating in the ocean would not result in any noticeable rise in sea level.

Fourth, the uplift and drawdown of the earth's crust can result in a rise or drop in sea level, primarily on the regional scale. Yet these processes generally take place extremely slowly. However, there are exceptions, as some considerable uplift or drawdown of the earth's crust has occurred over relatively short periods. Some parts of the Scandinavian Peninsula, having been freed from the ice masses of the last ice age, are rising by up to 1 cm per year.

Evaluations of sea levels measured over the 20th century show a rise of about 17 cm in the average sea level. This value was calculated by using the results of geological models of the movement of the earth's crust to correct the observed data to account for the effect of the uplift and drawdown of the earth's crust. Unfortunately, the uncertainty in the result is still quite great, because few long measurement series of the sea level are available, and because the corrections themselves contain uncertainties. Satellite measurements during the period 1993–2003 show a rise in sea levels of approximately 3.1 ± 0.7 mm/year, with thermal expansion contributing most and estimated at 1.6 ± 0.5 mm/year. The rise in sea levels thus appears to have accelerated somewhat in recent years.

Projections for the sea level over the 21st century are strongly correlated with the expected temperature rise. The climate models point to a further rise in sea level of 18 to 59 cm by 2100 (Figure 26). Yet in the extreme case this could be significantly higher, as models do not include future rapid dynamical changes in ice flow. How the large ice sheets will react remains an open question. However, if the concentrations of greenhouse gases continue to rise so strongly, it is entirely possible that considerable sections of the Greenland ice sheet will melt by 2100, thus adding yet another significant contribution to a rise in sea level. To quantify the contribution of the Greenland melt, a climate model was coupled with an ice sheet model. A simulation was performed in which the CO_2 content of the atmosphere was raised in a very short time from its preindustrial value of 280 ppm to a level about four times higher, 1100 ppm, and then stabilized at this value: the resulting increase in sea level was a maximum of 5 mm annually. After about 600 years, 40% of the Greenland ice sheet had already melted. It is therefore quite possible that the global sea level could rise by over 1 m by 2100, if the combined effects of thermal expansion and the melting of Greenland are considered in conjunction with some positive feedback mechanisms. Recent studies show that during the last major interglacial period, the Eemian interglacial around 125 000 years ago, sea level was about 4 m higher than it is today. Forecasts predict that the temperatures predominant during the Eemian will be exceeded in the Arctic by the middle of this century. However, for the record it must be stated that because the processes connected with the dynamics of ice sheets are not yet very well understood, any statements about the stability of the ice sheets entail high uncertainties (see section 1.4).

Furthermore, the rise in sea level can vary from one region to the next due to changes in the ocean currents. A collapse of the thermohaline circulation alone could make the sea level in the North Atlantic rise by up to 1 m, while sea level would drop in the south Atlantic. This has been simulated using oceanic models. In the equatorial Pacific, too, an El Niño-type change in circulation could result in considerable regional changes in sea level. In this case the sea level of the western Pacific could fall by as much as 20 cm and rise by a similar amount in the East Pacific.

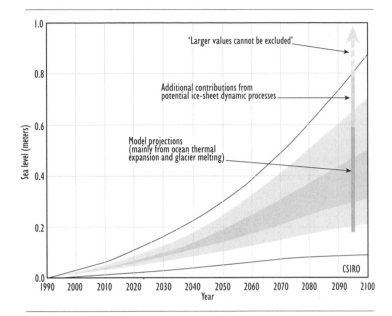

Figure 26 Projected sea-level rise for the 21st century: the range from
IPCC 2001 for the period 1990 to 2100 is shown by the lines
and shading

The central dark shading is an average of complex models for the range of SRES
emission scenarios. The light shading is the range for all models and all
scenarios and the outer bold lines include an allowance for land-ice uncertainty.
The IPCC 2007 projections are shown by the bars plotted at 2095. The lower
part of the bar is the range of model projections. Ocean thermal expansion and
melting of glaciers and ice caps are the largest contribution to this range. The
middle part of the bar is a potential additional contribution from a dynamic
response of the Greenland and Antarctic ice sheets to global warming (redrawn
after www.cmar.csiro.au/sealevel/sl_proj_21st.html)

Because of the great inertia of the oceans, the sea levels in
many of these predictions continue to rise even many centuries
after the concentration of CO_2 has stabilized (Figure 22). This
means, for instance, that sea level would rise by considerably

more than 1 m as a result of thermal expansion if the concentration of CO_2 were stabilized at approximately 1000 ppm in the year 2100, whereby the predominant rise in sea level would not take place until after 2100. Model calculations show that even after a thousand years, in the year 3000, sea level would continue to rise as a consequence of thermal expansion; by that time a rise of well over 2 m would be expected. Similar considerations are true for changes in the ice sheets of Greenland and the Antarctic. Many generations that follow us will be affected by the climate change we initiated. The rise in sea level could become one of the greatest problems facing mankind.

6.5 The Acidification of the Oceans

The oceans of the world store around 50 times more CO_2 than the atmosphere, and 20 times more than the terrestrial biosphere and soils (see Figure 9). Yet the ocean is not only an important CO_2 warehouse – in the long term it is also the most important CO_2 sink. Driven by the difference in partial pressure of CO_2 between the atmosphere and sea water, part of the CO_2 produced by human activity ends up in the surface layer of the ocean and, carried by the ocean currents over time periods from decades to centuries, ultimately in the deep ocean as well. An increase in the CO_2 concentration in the upper layers of the ocean has already been detectable for several decades, which can be traced back to the increased CO_2 content in the atmosphere. At present the ocean absorbs about 2 Gt of C every year, which amounts to about 30% of the anthropogenic CO_2 emissions from burning fossil fuels. In total the oceans absorbed about 118 ± 19 Gt C between 1800 and 1995, which amounts to about 48% of the CO_2 emissions from fossil energy sources (including cement production), or 27 – 34% of the total anthropogenic CO_2 emissions during this period, including those from changes in land usage. The anthropogenic CO_2 signal in the ocean is detectable down to an average water depth of about 1000 m; through the slow exchange of the ocean layers it has not yet reached the deep sea in large sectors of the ocean. Due to the formation of deep water (convection) in the North Atlantic, however, there the anthropogenic CO_2 signal extends to a depth of 3000 m. In the atmosphere CO_2 is for the most part chemically neutral: this means that it does not react with other gases, but makes its significant contribution to global climate change through strong interactions with infrared radiation. By contrast, CO_2 is chemically active in the ocean. Dissolved CO_2 decreases the pH value of seawater, leading to an acidification of the ocean.

This effect is already measurable: since the beginning of industrialization the pH value of the surface water of the oceans has dropped by an average of 0.11 units. Proceeding from a slightly alkaline preindustrial pH value of 8.18, the acid content on the surface of the ocean has thus increased. On the basis of various IPCC emission scenarios, an atmospheric CO_2 concentration of 650 ppm by the year 2100 is expected to cause the mean pH value to fall by a total of 0.30 units compared with the preindustrial value. At an atmospheric concentration of 970 ppm, the pH value would diminish by 0.46 units. If we succeed in limiting the amount of CO_2 in the atmosphere to 450 ppm, the pH reduction would amount to just 0.17 units.

The acidification of the oceans is an effect that can be traced back exclusively to the rise of CO_2 concentrations in the atmosphere. This distinguishes it from the climate change induced by the radiation effects of the increased CO_2 content, but also from the rise in methane, nitrous oxide and several other gases with strong climate effects. Discussions of climate change calculations thus often refer to the 'CO$_2$ equivalent,' whereby the radiation effects of the various gases are converted into the corresponding radiation effect of CO$_2$. The climate protection argument is that it does not make any difference whether the radiation effect is caused by CO$_2$ alone or by other emitted greenhouse gases. Yet this is not true for the contribution this gas makes to the acidification of the ocean: to protect the world's oceans, the only emissions that must be reduced are those of CO$_2$.

Acidification is primarily a consequence of the rapid increase of the amount of CO$_2$ in the ocean. When CO$_2$ enters the ocean slowly, it mixes all the way into the deep sea, where a gradual dissolution of calcareous sediment compensates for acidification. In this case the pH value of the ocean remains nearly constant. Slow inputs of CO$_2$ have occurred repeatedly over the history of the Earth, like at the end of the last ice age, when the concentration of CO$_2$ rose by 80 ppm over a period of about 6000 years, and in the climate epochs with high CO$_2$ content like the one approximately 100 – 200 million years ago. However, carbon dioxide enters the ocean because of us humans at least a thousand times faster than it does naturally.

The ocean is the most important net sink for atmospheric CO$_2$. Without the absorption of anthropogenic CO$_2$ by the ocean the CO$_2$ concentration in the atmosphere would be around 55 ppm over the present level, i.e., already considerably more than 400 ppm. By the way, the ocean is not the most important sink for the other greenhouse gases regulated in the Kyoto Protocol: the strongest sink for methane (CH$_4$) is the chemical reaction with the hydroxyl radical (OH) in the lower atmosphere, and most nitrous oxide (N$_2$O) is destroyed by solar UV radiation in the stratosphere. The ocean is an important source for N$_2$O, however, and it remains unclear how the content of this gas will fare under the impact of climate change.

The strength of the ocean as a CO_2 sink depends on how strongly anthropogenic CO_2 emissions are reflected in a rise of atmospheric CO_2 concentrations. The decisive factor for the ocean's ability to act as a CO_2 sink is thus the increase of CO_2 in the atmosphere. Before industrialization the ocean was not a CO_2 sink. On the contrary, its surface emitted around 0.6 Gt C to the atmosphere each year, with the same amount of carbon from the terrestrial biosphere (and thus ultimately from the atmosphere) simultaneously entering the ocean in the form of river-borne organic material. This did not change the atmospheric share of CO_2, which remained constant over millennia at around 280 ppm.

On the millennial time scale, the world's oceans are mixed through once and the carbon absorbed at the surface is distributed evenly. Thus in the long term the ocean will be able to absorb the majority of anthropogenic CO_2 emissions. Yet the more carbon is emitted, the smaller the share that can be absorbed by the ocean. And since the surface layers absorb carbon only until the partial pressures in the surface water and atmosphere reach equilibrium, in the short term, i.e., in the next decades and centuries, only part of this great sink potential will be realized. The limiting factor is the transport of the absorbed carbon material into the deeper ocean layers. Until now the oceans have absorbed only 30% of the anthropogenic carbon uptake that would be possible in the long term if atmospheric concentrations remain constant. In the last century the oceans' annual CO_2 uptake has risen nearly linearly with the atmospheric CO_2 concentration (0.027 ± 0.008 Gt C per ppm and year). In the future, though, a drop in the relative CO_2 uptake is expected: in addition to the emission path and the atmospheric CO_2 concentration, the future strength of the CO_2 sink will be determined by the capacity of the ocean surface layer to absorb CO_2, the speed of the exchange between the surface and deeper ocean layers, and the strength of the 'biological pump.' This last expression designates the process in which CO_2 is taken in by ocean organisms via photosynthesis and integrated into organic substances; dying organisms sink to the depths, which removes the carbon from the surface layer. Both climate change and CO_2 input affect these three processes and thus cause feedback effects that influence the efficiency of the oceanic carbon sink.

The chemical feedback mechanism through CO_2 input is well understood: the more CO_2 has entered the ocean already, the lower the carbonate concentration in the surface layer. This reduces the absorption capacity for additional carbon. If the concentration of CO_2 in the atmosphere rises by 100 ppm today (from 380 ppm to 480 ppm), the associated increase in carbon dissolved in the surface water is already 40% less than it was at the start of industrialization (for a rise from 280 ppm to 380 ppm). Even then it is still 60% higher than it will be in the future if atmospheric CO_2 rises, for example, from 750 ppm to 850 ppm. Even in the long term, on time scales of many centuries up to the millennia needed to mix through the ocean water thoroughly, this feedback mechanism has the result that the greater the total emissions of CO_2, the greater the share remaining in the atmosphere.

As temperatures rise, the solubility of CO_2 in ocean water decreases. Should ocean surface temperatures continue to increase as a consequence of the accelerating greenhouse effect, CO_2 absorption in the ocean surface layer will diminish. The increase in ocean surface temperatures also results in a more stable stratification of ocean water, such that the water is less well mixed with the water masses located below. Through this, and through the potential weakening of the thermohaline circulation, the vertical transport of ocean water enriched with CO_2

from the surface layer to the deep ocean is reduced, and thus the sink effect of the ocean.

An increased stratification (stability) would have a number of complex effects on marine biology. For one, the ventilation of the deep ocean and thus the influx of oxygen are reduced. Moreover, a more stable stratification decreases the return transport of nutrients to the surface layer and thus – by limiting the nutrients of primary production – can reduce biological productivity. At the same time, however, the return transport of carbon to the surface layer is reduced, so that even if biological productivity remains unchanged, more CO_2 is absorbed from the atmosphere. Depending on which of the two effects dominates, this can increase or decrease the sink function of the ocean. Carbon export by the biological pump also can change as a consequence of climate change and the acidification of the oceans. Different effects come into play here, ranging from a change in the nutrient networks all the way to a reduced rate of calcification, all of which require more intensive study before they can be modeled.

Although many of the mentioned effects (as well as other feedback mechanisms not listed here) are difficult to quantify, all of this suggests that in sum they contribute to a considerable weakening of the ocean's efficiency as a carbon sink. Various model results suggest that, by the end of this century, climate-conditioned feedback alone could cause cumulative CO_2 absorption by the ocean to end up $4 - 15\%$ lower than it would be without this feedback. These climate feedback effects are compounded by geochemical effects, which also contribute to a weakening of the relative sink. The greatest factor for uncertainty in estimating the future development of the ocean sink is the biological processes, that is, the effects on primary production, the biological pump and calcification. In spite of major, persisting gaps in our knowledge, it can be summarized that should atmospheric CO_2

concentrations continue to rise, the relative share of anthropogenic CO_2 emissions absorbed by the ocean will drop, even if absolute uptake continues to rise.

The anthropogenic increase in atmospheric CO_2 concentrations leads to shifts in the carbonate system of ocean water and to a lower pH value, and thus to the acidification of the ocean. The speed of the change currently taking place in the marine carbonate system is extraordinarily high, faster than has been experienced in the last 20 million years of the Earth's history. Man is interfering substantially in the chemical equilibrium of the ocean, and this will have consequences for marine life forms and ecological systems. A strongly increased CO_2 concentration has many negative physiological effects, which have been investigated experimentally on various marine organisms. Many changes in marine organisms have been demonstrated, for instance, in the productivity of algae, the metabolic rates of zooplankton and fish, the oxygen supply of squid, the reproduction of mussels, nitrification by microorganisms and the absorption of metals. From today's vantage point it is improbable that marine organisms will suffer from acute symptoms of poisoning due to the atmospheric CO_2 concentrations expected in the future. Yet for many species of phytoplankton, doubling the concentration of CO_2 increases the rate of photosynthesis by about 10%. However, the connections between photosynthesis, the primary production of phytoplankton and the repercussions in the nutrient network are complicated by a multitude of other factors (light and nutrient supply, differing risk of being eaten by zooplankton, adaptation processes, etc.), so that the effects of acidification on the growth and composition of phytoplankton are quite uncertain.

Besides photosynthesis, calcification is probably the most important physiological process affected by the increase in the concentration of CO_2, because it has far-reaching consequences for the ecological function of the marine ecosystem. As such, it also can trigger feedback effects on the concentration of CO_2 in the atmosphere, and thus on the climate system. Many marine organisms make their skeletons or shell structures out of calcium carbonate, which has to be extracted from ocean water. This is possible only when ocean water is oversaturated with calcium carbonate, which is why the drop in pH value as a consequence of rising CO_2 content makes calcification more difficult. The result is weaker skeleton structures, or – if the saturation concentration for calcium carbonate is no longer maintained – even their dissolution. Acidification has an impact on all marine species that calcify, including certain groups of plankton, mussels, sea snails and corals. Echinoderms like starfish and sea cucumbers are particularly at risk. Corals may be the most striking and best known calcifying marine organisms; yet while they suffer particularly detrimental effects under acidification, they contribute to only about 10% of global marine calcification. About three quarters of global marine calcification is caused by plankton organisms. Among these the coccolithophorides are of particular importance, since these protozoic primary producers, which can generate extensive plankton blooms with just a few species, make an important contribution to the export of calcium carbonate to the deep sea and thus play an essential role in the global carbon cycle. Experiments have demonstrated that the calcification of coccolithophorides drops off dramatically when atmospheric CO_2 concentrations rise.

The predicted reduction in pH values over the course of this century thus may have considerable effects on calcifying organisms and, consequently, on the marine biosphere as a whole. Parallel to this, a considerable increase in temperatures is expected due to the climate. These two effects are not independent of each other: for example, the rise in CO_2 can reduce animals' temperature tolerance. The ecosystems of the corals are the chief example of such negative effects. However, acidification could conceivably have consequences for the nutrient network as well. Different reactions to increased CO_2 concentrations could change the spatial and temporal distribution of species through the growth or reproduction of organisms due to changed competition. The acidification of the oceans could potentially have effects on fishing as well. Acute toxic effects of the increased atmospheric CO_2 concentrations on fish are not to be expected, since no acute sensitivity of these organisms to CO_2 will set in until concentrations beyond those forecasted. Yet what could become relevant for the fishing industry are those changes in the structure and function of marine ecosystems that are triggered by changes in the composition of species of phytoplankton due to hampered calcification, and which, through trophic coupling, have effects extending into the upper layers of the nutrient network. Moreover, the changes in the conditions for species' growth and competition in tropical coral reefs caused by acidification will have negative effects on fishing there. After the atmospheric effects and the increase in sea level, the acidification of the oceans is thus one of the greatest dangers anthropogenic carbon dioxide emissions present for the Earth system. However, many processes and feedback mechanisms in this extremely complex branch of research have yet to be investigated.

Strategies for the Future

7 The Public Discourse

7.1 The Role of the Media

Today we live in a media society. The highest good in such a society is attention. Because of the plethora of media (publicly and privately owned television and radio stations, newspapers, journals, magazines, the Internet) it is becoming increasingly difficult for journalists to win the attention of the viewer, the listener, the reader, or the Internet user. At the same time, it is ever more difficult for citizens to filter out the 'right' information from the flood of that proffered, whereby this is not the place to define which information is 'right.' The result of this development, for the climate debate at least, is a degree of uncertainty among the population. This uncertainty is partly responsible for the fact that climate protection measures are difficult to implement in society. On the other hand, without the media climate problems would not have reached the top of the global political agenda. In so far the media have achieved something that climate research could never have managed itself. The continuous reporting in the media thus deserves an overall positive assessment from the perspective of climate research.

Nevertheless some of the grievances about media reporting are worth pointing out here. In order to get citizens' attention, journalists often resort to the stylistic device of hyperbole. A prominent example of this is the title picture of the magazine *Der Spiegel* from summer 1986, which showed a photomontage of floodwaters reaching halfway up the Cologne Cathedral. A more recent example is Roland Emmerich's film *The Day After Tomorrow*. Although the consensus in science is that global warming cannot lead to a new ice age, this plainly incorrect scenario is described in the film and in many other media. But occasionally the device of playing down the issue is used as well, in order to attract the greatest attention possible. Several journalists have called the climate problem into question in order to set themselves apart from the 'uniform mass,' to raise their 'print run.' Indeed, it must be pretty boring to point out the climate problem over and again. Something new is needed to attract attention. What lends itself to this better than an 'expert controversy'? Accounts of this phenomenon abound in the media, even though it does not exist in the practice of climate research. This is not to say that there are no arguments among scientists. Yet the basic thesis of global warming as a consequence of the anthropogenic emission of certain trace gases into the atmosphere is accepted worldwide and beyond question.

The phenomenon known in Germany as 'Waldsterben' (forest dieback) is one example of the media's irresponsible treatment of a scientific subject. Although forestry scientists speak of 'Waldschäden' (damage to forests), in the media the forest problems are always discussed using the buzzword 'Waldsterben'. Damage to the forests has never been as serious as it is today, as proceeds clearly from annual reports on the subject. But today scientists face reproach because the forest is still alive after all. So even though the scientists' prediction was right, the media discuss the issue as if they had made an error. Some accounts draw the erroneous conclusion that the findings from all of the fields of environmental science, including those of climate research, cannot be trusted. This example illustrates clearly how exaggerations and an apocalyptic choice of words on the part of the media can discredit an entire branch of research. A potential result of this is a tremendous loss of credibility for the environmental sciences. A similar phenomenon can be observed in the public discussion about the climate problem. The media are fond of covering not the 'climate problem' or 'climate change,' but the 'climate catastrophe.' Because of this choice of terms, some contemporary media declare the predictions of climate researchers to be wrong: obviously, no catastrophe has taken place yet. But no serious climate scientist would ever speak of a 'climate catastrophe.' Nevertheless, if the calculations of scientists' models are compared with the actual trends in recent decades, they correspond quite strongly with scientific observations. The climate models have thus proven to be quite plausible, but this is not often the message communicated.

So what must be expected of journalists? The seriousness of the message must be in the foreground, i.e., neither the author's own opinion nor the compulsion to increase sales may distort the message so far that it becomes ambiguous or wrong. The media play an important role in our society and therefore must accept their responsibility to report as objectively as possible. Detailed background research is necessary, but hardly to be expected in today's fast-paced world, as it brings no rewards. Further, the media must not succumb to the fallacy that science can be discussed in the public sphere. The climate is an especially complex physical system, which is described in terms of extremely difficult mathematical equations. The scientific questions therefore must be discussed in the scientific sphere, and the role of the media restricted to publicizing the sustainable results of scientific research. Naturally, this also includes the task of communicating where knowledge gaps still exist.

The role of experts also deserves critical attention, for a variety of reasons. Most scientists have never learned to present their results such that they can be understood by laymen, but most journalists are not scientific experts. Misunderstandings are thus bound to occur. Moreover, scientists have to keep in mind that laymen pick up and understand the language of science differently than do scientists themselves. A subjunctive construction, for instance, often goes unnoticed by the public, just as statements about probability are generally misunderstood, completely or at least in part. Thus experts must take great care in their formulations. This costs more time than many experts are willing to invest. Like the journalist, the expert, too, must not pursue any agenda of his or her own, for instance, by placing certain aspects in the foreground in order to promote his or her own agenda. This point is especially important, for every true-blooded scientist tends to overestimate the importance of his or her own work. Just because a process, let us say cloud formation, does not receive accurate consideration in the models, this does not mean that the models are worthless. So the cloud physicists serving as our example here must therefore carefully consider how to communicate these uncertainties to the public without calling into question the results of the climate models in general and the climate problem as a whole.

Where are we in the public climate debate today? A diffuse picture of the causes and effects of global climate change predominates. Complexes of issues like the ozone hole and the greenhouse effect are often confounded, by the media – even in broadcasts like the nightly news on the state-owned stations – and also by the population. Apparently it does not seem possible to establish a solid basis of knowledge in society. Yet there is a clear consciousness in the public that the climate has already changed as a consequence of human activity, and that it will continue to change. Yet most people are not clear about how the human influence on the climate manifests itself in detail.

For the relationship between journalists and experts it would be desirable if a kind of solidarity were to emerge between the two camps. On the part of science there should be insight that media work is not merely a necessary evil, but actually an important basic requirement for science to become anchored in society. The process of shaping public opinion on policy requires science. Experts must be willing to accommodate journalists to some degree, by simplifying complex issues in order to make them accessible to the public. On the other hand, journalists should show some understanding for the fact that certain scientific issues cannot be simplified at will, and that some complexes of scientific questions are not suitable for sensationalist reporting. Journalists and experts are in the same boat; they depend on each other. Only if both groups are satisfied with the state of the public discussion can it be presumed that the population is competently informed about the state of affairs on global climate change.

7.2 The Skeptics

What is published about global climate change is becoming increasingly contradictory: some authors warn about a 'Hot Age,' while the others see the next ice age approaching. Some speak of melting poles, the others of freezing poles; some warn of increasing storms, flooding, deserts, the others expect paradise on Earth; some fear a host of climate refugees from the South, others fear the impoverishment of the North because of excessive expenditures on climate protection; some are for immediate action, others advocate a wait-and-see strategy or believe everything to be a ploy by climate scientists craving validation and thirsting for research funds. It is difficult for the layman to get a well-founded idea of the situation. The material is difficult, access to original literature tedious, and what has been published is often incomprehensible to anyone but experts. Compounding this is media coverage that is designed to be as up-to-date as possible, which practically invites unreflected reporting. Against this background it is quite easy for so-called 'skeptics' to attract attention. Their claims include the discovery of a crucial new process that relativizes human influence on the climate, or the inability of climate models to simulate elementary properties of the climate, which robs their predictions of any credibility. The skeptics' objective is to make light of the climate problem and prevent the institution of measures to protect the climate.

At this juncture it is therefore necessary to address the popular arguments and assertions of the skeptics who see no need for action as regards climate change. A particularly prominent example of a climate change critic is the writer Michael Crichton (*Jurassic Park*), who asserts in the novel *State of Fear* that climate scientists essentially made up the climate problem and entered into an unholy alliance with the media. Reading this book makes all too clear that Crichton has fallen for the popular skeptics' arguments. Some of these arguments are plausible at first glance, but they do not stand up to scientific scrutiny. Some can be described as spurious arguments, which sound good but possess no scientific validity whatsoever. Often they are plainly wrong. Moreover, these arguments usually come from people who are either not active in climate research or have no training in the natural sciences. In so far it is worth taking an occasional look at a skeptic's resumé. More often than not, such a glance will reveal ties or even intimacy with the mineral oil lobby. The following section will examine several of these arguments to see if they hold water. Our investigation here will be restricted to several of the most frequently advanced arguments, which were thematized in a brochure by the German Environmental Agency (Umweltbundesamt 2004).

Climate cannot be predicted in the long term because weather forecasts are possible only several days in advance

One of the most common arguments by skeptics is: how can it be that climate prognoses are possible for the next hundred years, when weather can be calculated only several days in advance? The difference between weather and climate, or the issue of climate predictability, was dealt with above (see section 4.4). At this juncture let us describe an example that makes clear how absurd this argument really is. Each of us can make a climate prediction, and do so in such a way that we predict that the coming summer will be warmer than the past winter. This forecast may sound trivial at first. But if the skeptic's arguments are taken to their logical conclusion, not even this trivial prediction should be possible, as it covers a time period clearly longer than several days. This forecast, as simple as it is, can be made, as we all know, because the position of the sun changes. In mathematics this is referred to as a change in the 'boundary conditions.' So when boundary conditions change, climate prognoses are possible. An important boundary condition for the climate is the chemical composition of the atmosphere. Since we are affecting this through our emissions of greenhouse gases, the climate has to change. Hence we can state with certainty that a necessary consequence of this emission is global warming (see sections 2.2 and 2.3).

In this sense, the prediction of global climate change is just as simple as the trivial summer-winter example above, at least when the Earth is regarded globally. For this reason the first predictions of global climate change are already over a hundred years old.

In so far chaos and predictability are not mutually exclusive. The only short-term possibility of weather forecasting is conditioned by the somewhat chaotic nature of weather events. Here the term 'chaotic' means simply that a minor change in the initial constellation can trigger ever greater differences in what happens over the course of events. Yet this does not mean that chaotic systems are not predictable under certain circumstances. A further example is the simultaneity of microscopic chaos and macroscopic conformity to physical laws in the behavior of gases. Consistent with kinetic gas theory, individual molecules execute chaotic, hardly predictable courses of movement, yet the external conditions of pressure, volume and temperature assume well-defined values and are linked with each other by means of strict physical laws.

Climatology has laws and connections analogous to this, which make it possible to describe long-term evolution, especially when boundary conditions change in a predictable way. Physical laws, in the form of mathematic equations, then allow statements to be made about the statistics of weather phenomena – such as the future frequency of storms – although not about the occurrence of individual events. This is similar to our example of the loaded die. We know that six will come up more often, but we do not know what the next roll will be.

There is no such thing as the greenhouse effect

Another argument by the skeptics refers to the natural greenhouse effect, the very existence of which is denied. The thesis of the greenhouse effect, they claim, violates the second fundamental theorem of thermodynamics, since no flow of energy is possible from a cold body to a warm one. Further, they claim, the hypothesis of the greenhouse effect must be incorrect for the simple reasons that the Earth is not a closed system and there is no glass roof in the sky. In fact, the greenhouse effect is textbook knowledge. It was proven long ago, both theoretically and experimentally. What matters here is not the existence of a glass roof, of course, but that atmospheric trace gases absorb infrared (long wave) radiation.

That the greenhouse effect does in fact exist and function is common knowledge to anyone who has ever entered a car parked in the sun. It is warmer inside than outside – as if the heat had been separated out, which should not be possible according to the second fundamental theorem of thermodynamics. In reality, however, the process that takes place is driven by a strong application of energy in the form of visible light (short wave radiation) through the windows of the car. The higher temperature in the interior of the vehicle results from the prevention of the equal distribution of this applied energy, as glass is practically impenetrable for the heat generated through the conversion of light. Once the application of energy has ended, the required equilibrium is gradually restored. In the atmosphere the greenhouse gases assume the role of the glass, preventing the removal of the (thermal) energy. In order to restore the equilibrium between absorbed (light) energy and emitted heat, the radiating surface of the Earth must assume a higher temperature. In total – and for this the second fundamental theorem of thermodynamics applies – net energy flows from the warmer Earth's surface to the colder atmosphere, because the thermal radiation of the Earth's surface is greater than the back-radiation of the atmosphere. The second fundamental theorem is thus not called into question.

Occasionally some skeptics advance the argument that the human contribution to the greenhouse effect makes up only around 2% and is therefore irrelevant. The percentage is correct, yet 2% of the total greenhouse effect (33°C) corresponds to around 0.6°C. And this is approximately equal to the human share of global warming in the last hundred years.

Climate scientists' prognoses are constantly being corrected downward

Skeptics are fond of claiming that the temperatures predicted by climate scientists always turn out to have been too high. Far from being true, this impression is the result of comparing apples and oranges. The first IPCC report (1990) stated that should the CO_2 concentration double, the global rise in temperature would amount to $1.5 - 4.5°C$, whereby the large range had its source in the representation of various feedback processes (especially in the hydrological cycle and the way cloud processes are portrayed) in the different models. This range of what is known as 'climate sensitivity' (the change in temperature for a doubling of the CO_2 concentration) remains unchanged even today.

In 1995, 2001 and 2007 the IPCC then made statements about possible temperature increases up to the end of the 21st century. However, an entirely different complex of questions was involved in these statements. This rise in temperatures also depends on the trend of emissions, and thus on human behavior during this period, yielding a correspondingly expanded range of temperatures. In the second IPCC report (1995) the temperature increase by 2100 was given as $1.5 - 3.5°C$; in the third assessment report (2001) the range is estimated to be $1.4 - 5.8°C$ and in the last assessment report of 2007, $1.1–6.4°C$. The temperature values in the second report are numerically lower than those in the first, which some readers interpreted as an all-clear signal lifting the alarm raised by its predecessor. But in this case climate sensitivity, which is a property of the model, is compared with a potential change in temperature by 2100, that is, with calculated scenarios. These are two fundamentally different variables. In so far there is absolutely no evidence that the prognoses for the Earth's warming have been corrected downward. In fact, the climate sensitivity of the climate models has barely changed at all in the last 20 years, although they are taking ever more processes into consideration. The IPCC report of 2007, for instance, gives a best estimate of the climate sensitivity of 3°C. Thus it is important to establish precisely what kind of statement is being advanced. The future development of the climate depends especially on which scenario is presumed, that is, on how the emissions of greenhouse gases develop in the future. The wide range of 1.1 – 6.4°C by 2100, which was given in the last IPCC assessment report of 2007, documents above all that we do not know today how we will behave in the future. Warming will turn out to be moderate if we sink greenhouse emissions strongly; it will be strong if emissions continue to grow.

Warming is a good thing

A slight temperature rise can indeed mean better growing conditions for the agriculture and forests of the moderate latitudes. What is important about this is how future rainfalls will be affected. A rise in temperature can mean significant problems for warmer countries (in the Mediterranean region, for instance) where droughts are increasing even now, and also in the Arctic regions, where the permafrost is about to melt, already posing serious problems for the region's infrastructure (buildings, streets, pipelines, etc.). This brings us back to the intensification of tropical cyclones like hurricanes, which has been under way for around 20 years and has led to tremendous damage. The intensification of tropical cyclones worldwide may already be a consequence of the moderate warming of the tropical oceans by a just a few tenths of a degree.

In our latitudes, too, even a moderate temperature increase is linked with negative consequences, like the melting of glaciers, the migration of species, heat stress, longer droughts and more strong rainfall. For instance, the frequency of strong rainfall events over most land areas has already increased in the last hundred years. Yet all indications point to the fact that, without countermeasures, the temperature rise we will be facing will be not moderate, but rather considerable, and probably associated with a significant increase in extreme events. Thus there is no occasion to speak of a desirable improvement in our situation. According to all calculations, the negative consequences of the anthropogenic climate change will outweigh the positive effects, such that it is important to keep the Earth from warming up too much. Especially important to mention here is the rise in sea level, which could well amount to 1 m by 2100. Over the next thousand years, sea level could rise by many meters.

We are already on our way to the next ice age

This thesis proceeds from the presumption that the global climate is governed not by a change in the chemical composition of the atmosphere, but by the parameters of the Earth's orbit. Variations in these parameters are doubtlessly an important cause of climate change, but by no means the only one. Climate change is not a monocausal event; various factors (orbit parameters, the intensity of solar radiation, the anthropogenic greenhouse effect, changes in land usage, etc.) operate simultaneously. However, the time scales on which these factors exert effects are highly differentiated. The cycle of ice ages and interglacials, in particular, unfolds over periods lasting many millennia. Thus the climax of the last ice age was around 20 000 years ago. In so far the extremely long processes that cause ice ages remain irrelevant for the next hundred years; thus we need not have any qualms about neglecting them in our consideration of anthropogenic global climate change.

On the basis of findings in the field of paleoclimatology, especially from the analyses of seafloor samples (sediment cores) from the deep sea and ice cores from the Antarctic, by now it is considered certain that the changes in the Earth's orbit are the main pacesetter for the ice age-interglacial cycles. These orbit parameters change in cycles around 20 000, 40 000 and 100 000 years in duration (see section 3.5). The overlapping of these cycles allows us to calculate that we are still around 50 000 years away from the anticipated climax of the next ice age. Climate changes based on the greenhouse gases released by man, on the other hand, will take place in a considerably shorter time frame, fifty to one hundred years, and thus at a much higher speed. Therefore it is illusory to hope that the global warming initiated by us humans could be compensated for by an imminent ice age.

The sun influences the climate, man is innocent

The sun is the only energy supplier for the Earth's climate system. Therefore it is a matter of course that changes in the radiant intensity of the sun, i.e. changes in the influx of energy, exert direct effects on the climate. The radiation of the sun is not as constant as was originally thought. It has been known for a long time that the number of sunspots fluctuates in cycles lasting eleven years, more or less. The intensity of solar radiation is also linked with this cycle. The concrete duration of the cycle (between eight and seventeen years) may also serve as a measure of the given radiant intensity. Since the late 1970s it has been possible to observe solar radiation directly from satellites. Such observations have shown that the difference between the maximum and minimum of solar radiation during a sunspot cycle amounts to about 0.1% of the radiant intensity, which is too little to have any significant influence on the surface climate (see section 5.3). So far no superimposed increasing or decreasing trend has become apparent.

Another amplifying mechanism is often evoked, according to which solar irradiation indirectly influences cloud cover:

the more intensively the sun shines, so the assertion, the more marked the interplanetary magnetic field becomes. This field serves to deviate cosmic radiation and thus prevents it from reaching the Earth's atmosphere. Cosmic radiation, the argument continues, results in the formation of condensation nuclei, thus facilitating cloud formation. Certain of these clouds would then provide shadow, and thus lower temperatures; accordingly higher radiation activity would result in less clouding and thus relatively higher temperatures, and vice versa. This argumentation is deductively plausible, but it is not clear how important this process is in relation to the other processes involved in cloud formation. In particular, according to today's meteorological state of knowledge, the presence of condensation nuclei is not necessarily a triggering factor in cloud formation. Whether additional condensation nuclei lead to more cloud formation cannot be assessed conclusively, because in any case they are present in sufficient number. A further question is whether this process would create primarily low cumulus clouds, which have a cooling effect, or higher cirrostratus, which make a net contribution to the greenhouse effect and thus cause warming (especially at night).

The first factor that supports the influence of the sun is a documented high correlation between the length of the solar cycle (as a measure of radiant intensity) and the global mean temperature for the period from 1850 to 1980, and between the intensity of cosmic radiation and the degree of cloud cover for the period from 1984 to 1993. However, the correlation between the duration of the solar cycle and the temperature resulted from an inadmissible statistical treatment of the data and was thus withdrawn by its originator. By contrast, the correlation established between cosmic radiation and cloud cover was apparently of a coincidental nature, as it has not been confirmed over the passage of time. Furthermore, the cosmic radiation dropped during the last 20 years while the temperature strongly increased.

Worth mentioning here is that existing correlations are not sufficient proof of causal relationships: they may be traced back to shared causes in a third factor, or even be entirely coincidental. The best known example is the correlation between the birth rate and the number of storks. As such it is important that any statistical connection, such as a relatively high correlation, be accompanied by a comprehensible physical theory that can explain the discovered statistical relationship.

Another popular assertion by skeptics is that changing cosmic radiation is the real reason for the change in the Earth's climate, although this process takes place over hundreds of millennia. Aside from the fact that the paper in question was subject to extensive critique for its methodological deficiencies, even if the argument were true it would be entirely irrelevant for the development of the climate in the next decades or centuries. Here we recognize another recurrent argumentation pattern: the impermissible practice of invoking processes that may operate on extremely long time scales, but have no meaning at all for the short time scales on which anthropogenic climate change is taking place.

Even if the amplification mechanism mentioned above were valid, on the basis of the direct observation of the intensity of solar radiation over the last 25 years no sustained trend would be expected. The only pattern revealed in solar radiation resembles a cyclical see-saw. In reality, however, we observe a temperature rise of around 0.2° per decade over the last 25 years, which amounts to a quite marked, sustained warming trend (see section 5.3). Beyond this we cannot recognize any appreciable eleven-year cycle in the temperature near the surface, so that its influence on the climate of the lower atmosphere must be minimal.

The influence of solar fluctuations in the upper atmosphere – in the stratosphere, for example – can be quite considerable, and is often advanced as proof for the climate efficacy of changes in solar radiation. However, it is important to differentiate between the upper atmosphere and the Earth's surface. The discussion about the anthropogenic greenhouse effect concerns chiefly the warming of the Earth's surface and the lower air layers, and obviously these have hardly been influenced at all by fluctuations in the supply of solar radiation in recent decades.

This does not mean that the sun has never exerted effects on climate conditions in previous periods and over longer time frames. During what is known as the Maunder Minimum (1650 –1710), for example, very few sunspots occurred, and temperatures were very low, which is why this period is referred to as the Little Ice Age. The connection between the climate during the Little Ice Age and the sun has since been confirmed by climate model simulations. Yet statistical analyses and model calculations for the last 25 years yield quite a strong dominance of the anthropogenic greenhouse effect as the cause of the rise in temperatures. Moreover, no scientific work has been submitted that manages to explain the strong warming of the last decades without taking into consideration the rapid rise in the atmospheric concentration of greenhouse gases.

Water vapor makes a significant contribution to the greenhouse effect

There is no question that by far the greatest contribution to the natural greenhouse effect is made by water vapor. Around two thirds of the natural greenhouse effect can be attributed to this gas. From this skeptics are fond of jumping to the conclusion that water vapor plays a greater role than carbon dioxide in the anthropogenic greenhouse effect. For starters, the natural greenhouse effect must be distinguished from the anthropogenic one. While water vapor is the most important gas for the natural greenhouse effect, in the anthropogenic effect carbon dioxide plays the largest role, contributing about 60%.

The emission of water vapor on the Earth has practically no effect at all on the greenhouse effect caused by humans, because this does not cause any permanent increase in the concentration of water vapor in the atmosphere. While carbon dioxide can remain in the atmosphere for approximately one hundred years (see Table 1), water vapor generally returns to the Earth in the form of precipitation after just a few days. A permanent increase in the content of water vapor does occur, however, when a rise in global temperatures causes evaporation to increase on the one hand, and on the other increases the potential capacity of the atmosphere to take in water vapor. The warmer the atmosphere is, the more disproportionately water vapor can be held in the atmosphere before condensation begins. This circumstance is described by what is known as the Clausius-Clapeyron equation. Water vapor is thus an important feedback gas, and positive water vapor feedback is the most effective of the various feedback processes. A warming induced by a rise in the concentration of carbon dioxide, for instance, leads to more water vapor in the atmosphere, further amplifying the greenhouse effect. In spite of several media reports to the contrary, water vapor is also taken into consideration in model calculations as a matter of course, and it makes up a considerable share of the total resulting rise in temperature in response to enhanced carbon dioxide concentrations.

Carbon dioxide cannot be the catalyst of climate change: climate history tells a different story

When the trends in the carbon dioxide content of the atmosphere are compared with the mean temperatures over the last 800 000 years, the high degree of correspondence (parallelism)

of the two cannot be overlooked (see Figure 8). From this isolated conclusions have been (impermissibly) drawn in the past
that climate events in the past were driven by CO_2. In reality it is
not immediately clear whether the variables influence each other
causally, whether both are governed together by a third variable,
or whether the correspondence is purely coincidental.

More precise analysis of the trends shows that Antarctic
temperatures about 240 000 years before the present advanced
slightly before the concentration of CO_2, about 800 years earlier.
Thus these long-period pathways show a change in temperature first, primarily due to the changes in the parameters of the
Earth's orbit. The increase in temperature, through the rise in
the water temperature of the oceans, causes the release of dissolved CO_2. The decomposition of biomass may also react more
strongly to the temperature than does its composition. Through
this the increasing temperature also results in rising CO_2 content
in the atmosphere, further amplifying the greenhouse effect.
The phenomenon observed here, like water vapor feedback, is
thus a positive feedback mechanism. Like other feedback effects,
these amplify the relatively minor climate changes caused by
the changes in the parameters of the Earth's orbit so intensely
that massive climate fluctuations like ice ages and interglacials
result. The present situation is completely different. The orbital
parameters can be considered as constant, as they vary only on
very long timescales. Carbon dioxide is emitted by us humans
into the atmosphere, and this leads to global warming through
an enhanced greenhouse effect. The associated changes in the
carbon cycle will further amplify the initial warming. Because
of the strong interaction between physics and biogeochemistry,
coupled climate/carbon cycle models are becoming ever more
popular to calculate the climate effects of anthropogenic greenhouse gas emissions.

The concentration of carbon dioxide cannot double because there are not enough fossil stocks

This claim might have some validity for the fossil reserves secured to date, for while burning these reserves would double the carbon dioxide concentration over the preindustrial value of 280 ppm in the atmosphere, only about half would remain in the atmosphere (see section 2.4), as the other half is absorbed by the oceans and the terrestrial biosphere. Therefore it is true that it would not be possible to double the content of carbon dioxide in the atmosphere by burning the fuels available today.

Yet it must be kept in mind that the actual amounts of fossil fuels available are much greater than the stocks secured at this time (see section 8.3, Figure 29). For coal, for instance, thanks to new exploration and excavation technologies, around ten times the current amounts are expected; with further technical advances this value constantly has been and will continue to be corrected upward. Moreover, there are additional fossil fuels such as methane hydrates: solid, ice-like materials made of methane and water. They are stable only at low temperatures and high pressure, and are found on the ocean floor and in permafrost soils. In the long term, therefore, significantly higher quantities of fuels must be expected. How much methane hydrate there is on Earth is a matter of controversy. One estimate that is often cited quantifies deposits at 10 000 billion tons of carbon – double the amount of all coal, gas and oil deposits worldwide. Certainly these deposits would be sufficient to more than double the carbon dioxide content of the atmosphere. Other 'exotic' resources such as shale oil, bitumen, and heavy oil also exist. A CO_2 concentration of up to 4000 ppm within the next centuries is thus altogether possible.

It must also be kept in mind that only about 60% of the greenhouse effect caused by man is traced to carbon dioxide. Without any countermeasures, the 'equivalent' CO_2 content taking into account the other greenhouse gases would probably double well before the middle of this century. Moreover, it is difficult to predict whether, with a rising concentration of carbon dioxide and the expected warming, carbon dioxide can continue to be taken in by the oceans and vegetation of the future to the degree it is today. With rising temperatures and correspondingly strong heat stress, the biosphere that has served as a CO_2 sink so far might even become a source, releasing additional CO_2. Furthermore, it is generally accepted that the relative capacity of the oceans to absorb carbon dioxide will decrease in the future (see section 6.5).

Carbon dioxide comes primarily from oceans or volcanoes

The annual increase in CO_2 concentration (an average of 1.5 ppm/year, with considerable variation and a present rate of 2 ppm/year) and the trend of concentrations since the preindustrial era are well known and by now also uncontested. Isolated arguments are heard asserting that most of this CO_2 comes from the oceans, from which it escapes as a consequence of (natural) warming (as if from warmed fizzy water). The validity of this assertion can be checked experimentally, by measuring the carbon dioxide content in the atmosphere over the ocean, and in parallel, the amount of carbon dioxide dissolved in the water of the ocean. Doing so shows that the quantity of carbon dioxide dissolved in the water is generally less than the CO_2 contained in the atmosphere. The majority of the Earth's oceans are thus 'undersaturated' in terms of their carbon dioxide content. Therefore, in balance, carbon dioxide predominantly passes from the atmosphere into the oceans; all in all, the oceans function as a carbon dioxide sink. This fact is also confirmed by measuring for carbon isotopes. Measurements of this kind allow carbon dioxide from natural (biogenic) sources to be differentiated from that generated by the burning of fossil fuels (coal, petroleum, natural gas). Measurements of carbon isotopes confirm that the increase of carbon dioxide in the atmosphere can be traced back to the burning of fossil fuels, and not to gas released by the oceans. Further, it has been observed (see section 6.2) that the oceans are becoming increasingly acidic – more evidence for the fact that the oceans are taking in carbon dioxide. In the long term this acidification can have grave consequences for marine life. Should worldwide CO_2 emissions continue to rise strongly in the future, there is a real danger that the pH value of the oceans may reach a value not experienced for at least 20 million years.

To date, the measurements available have suggested that relatively minimal CO_2 emissions originate from volcanoes. Recently some voices have asserted that the amounts released by volcanoes are considerably larger, and that CO_2 is also emitted by rocks, primarily in volcanic areas. The quantities mentioned amounted to about 600 million tons of CO_2 per year, which would correspond to about 2% of today's anthropogenic emissions. Even if this quantity happened to be true – which is very difficult to establish using the measurement technology available today, this would have absolutely no effect on anthropogenic emissions. A relatively well known, large quantity of anthropogenic emissions would be increased by a poorly known minimal amount of natural emissions. A high rate of emission from volcanoes and rock is improbable for the simple reason that the CO_2 concentration in the atmosphere was relatively constant in the 800 years before industrialization. If volcanism had been the main emitter, the concentration would have had to fluctuate strongly in this period as well.

More carbon dioxide has no effect on the climate

It is often asserted that extra carbon dioxide introduced into the atmosphere has no relevance for the climate, since the essential absorption bands are already saturated. This argument is valid for several ranges of the spectrum (especially the central ranges of the absorption bands at wavelengths of 4.5 and 14.7 μm), but not for all ranges, and in particular, not for the flanks of the absorption bands. This circumstance means that CO_2 does indeed affect the climate, although, because of the high degree of saturation, it has much less global warming potential than materials that are fully absorbent (in unsaturated infrared bands). The importance of CO_2 thus results from the enormous quantities in which it is emitted. Because carbon dioxide is effective chiefly at the flanks of the absorption bands, its influence on the climate increases only with the logarithm of the carbon dioxide concentration. But because the concentration of CO_2 is rising exponentially, the rise in temperature should be more or less linear over time, which is also demonstrated in the models. Moreover, water vapor feedback and other positive feedback mechanisms lead to an amplification of the pure CO_2 effect and of the effects of the other greenhouse gases we humans introduce into the atmosphere.

Discrepancies exist between satellite measurements and ground measurements of temperatures

Skeptics long claimed that there was a contradiction between the trends of ground-level measurement stations, which recorded a rise of almost 0.8°C in the global mean temperature over the last hundred years, and the available satellite data, which have been measured for only the last 25 years, but indicate no trend or even a slight drop in temperature. Yet these records, obtained using various methods (ground and ship measurements are direct measurements of air temperature, while satellites measure the temperature indirectly via the long wave radiation emitted by the Earth's surface and the atmosphere), are comparable only to a limited degree. Reference was made above to the very different temporal dimensions of both measurement series. This also means that the experience in the collection of ground data is much greater and the development and standardization of measurement technology much further advanced than for satellite measurements, where, for example, very different measurement devices and procedures were and still are deployed. Particular difficulties are posed by the combination of data from different satellites, because they are separated by considerable gaps, making it practically impossible to detect trends. Also crucial are the differences in the variables of observation: ground-level measurements are point measurements at an altitude of 2 m, whereas satellites record a vertical integral, i.e. a vertical mean, of the temperature with the largest contribution from the lowest 6 km of the troposphere (in this altitude range, temperatures can extend over a bandwidth of over 30°C). For all of these reasons, the direct measurements are clearly more suitable for statements about a long lasting temperature trend on the Earth's surface than are the satellite data. Moreover, the ground-level temperatures are decisive with regard to effects on animate and inanimate nature.

It was demonstrated that the satellite data were calibrated incorrectly, because the model failed to take into consideration the satellite's loss in altitude of 1.2 km per year, which affects the measurement results. Since a correction was implemented, the satellite data, too, have shown a slightly rising trend (of approximately 0.05°C per decade). The contradiction dissolved completely after a recent publication ended up with practically the same temperature trends as the direct temperature measurements, using the very same satellite data but a different statistical method. The IPCC states in its 2007 report: 'New analyses of balloon-borne and satellite measurements of lower- and mid-tropospheric temperature show warming rates that are similar to those of the surface temperature record and are consistent within their respective uncertainties.'

Over all, the common skeptics' arguments thus do not hold much water. That is why there is such a strong consensus in international climate research that man is influencing the climate to an ever greater degree. The hard facts:

1. Greenhouse gas concentrations are rising massively as a consequence of human activity. The atmospheric content of carbon dioxide, for example, has not been as high as it is today for over 800 000 years.
2. This has led to an amplification of the greenhouse effect and to global warming. Of the total warming of the Earth observed in the last hundred years, about 0.6°C of the total 0.8°C are attributed to man. A lesser contribution to the warming of the Earth, around 0.2°C, is of natural origin.
3. Global warming has led to an increase in extreme weather events all over the world in recent decades.
4. Warming has caused the Earth's snow and ice cover to recede. The warming of the oceans and the melting of land ice has resulted in a sea level rise of about 20 cm. Should greenhouse gas concentrations continue to rise, the warming of the Earth by 2100 will be unique for humanity.
6. The consequences would be an additional increase in weather extremes and a further rise in sea level.

8 What must be done?

8.1 The Kyoto Protocol

At last the climate problem has made it to the top of the global political agenda. The United Nations Framework Convention on Climate Change of Rio de Janeiro in the year 1992 defined the goal of stabilizing atmospheric greenhouse gas concentrations on a level 'that would prevent dangerous anthropogenic interference with the climate system.' In December 1997, 159 signatories of the United Nations Framework Convention unanimously accepted what is known as the Kyoto Protocol. As demanded by the first conference of signatories in the Berlin Mandate of 1995, this protocol formulated the first implementation rule on the climate convention. It obligates industrialized countries to reduce their emissions of greenhouse gases by an average of 5.2% (based on emissions in the year 1990) by the period from 2008 to 2012. A five-year framework was chosen rather than a precise goal year to compensate for unusual fluctuations (for example, due to special weather conditions, such as extraordinarily cold winters). The longer obligation period also allows the framework to account for factors like carbon sinks and emissions trading. The reduction in emissions of greenhouse gases other than carbon dioxide can be offset by converting these emissions into CO_2 equivalents. The Kyoto Protocol will become binding under international law when it has been ratified by at least 55 of the signatories, which accounted for at least 55% of all CO_2 emissions in 1990. In February 2005 the Kyoto Protocol came into force upon its ratification by Russia.

The European Union must reduce emissions by an average of 8%. Yet there is burden sharing within the EU, with very different country-specific reduction objectives. The United States are supposed to reduce 7% and Japan 6% of their emissions of greenhouse gases. So far, however, the United States has not ratified the protocol. At present American voices are vehemently against the Kyoto Protocol. Russia was supposed to merely stabilize emissions, and Norway is even allowed to increase. These varying reduction rates are the result of demonstrably different conditions, but also in part a consequence of the negotiation skills of individual countries. No binding reduction goals at all were formulated for developing countries such as China and India.

With the Kyoto Protocol, now that human influence on the world's climate has become obvious, mankind is taking its first steps to manage the Earth system. The Kyoto Protocol contains four instruments that are new to mankind:

1. emissions trading
2. joint implementation
3. the calculation of net sources, i.e. the inclusion of emissions and sinks for greenhouse gases, often abbreviated as 'accounting for sinks,'
4. a 'clean development mechanism' for a sustainable economy.

Codes of practice have yet to be defined explicitly. For example, it must be clear what share of the reduction obligation a country may compensate for by purchasing emissions rights from another country that succeeded in reducing its emissions beyond the stipulated goals. Other definitions must be agreed upon, such as how much CO_2 is bound by a newly reforested wooded area and thus may be subtracted from the emissions caused by the consumption of fossil fuels. For the first time there is also joint implementation between Annex I states (the industrial countries listed in Annex I of the Convention) as part of a protocol binding under international law. Yet what happens, for example, if Germany builds a gas and steam turbine power plant in Ukraine, with an electric efficiency of nearly 50% and simultaneous waste heat utilization, on the location of an older power plant that had an electric efficiency of only 30% and no waste heat utilization? How much carbon dioxide savings result, how are these credits accounted for, and to which country? The rules will become even more complex when this mechanism for sustainable development is applied to technology transfers to developing countries. In such cases it must be ensured that the motivation for innovation in the highly developed countries remains strong, so that new technology will flow more quickly to the developing countries and receive the requisite financial support. At least some of the money required for such transfer could come from emissions trading, but for this, too, rules have yet to be defined.

Yet the Kyoto Protocol does not provide the climate protection needed from the perspective of climate researchers, and in so far its value is above all symbolic. Moreover, at this time the United States are not willing to ratify the Kyoto Protocol, although this country alone is responsible for about 20% of the (energy-related) CO_2 emissions worldwide, ranking number two on the world's hit list of the largest carbon dioxide emitters. China has recently climbed to the top of the hit list and is not even addressed in the binding measures of the Kyoto Protocol. In order to prevent severe climate changes in the next hundred years, the amount of greenhouse gases emitted today would have to be at least halved by the year 2050. This is far more than what is demanded in the Kyoto Protocol. The Kyoto Protocol therefore is just a first step, but could be an important one toward sinking the global emissions of greenhouse gases.

However, the annual conferences of signatory states present opportunities for amendments, as was the case for the Montreal Protocol, the regulatory statute on the Vienna Convention for the Protection of the Ozone Layer. The original Montreal Protocol of 1987 barely provided for protection of the ozone layer – some scientists on the team called it mere 'euthanasia' for the ozone layer. However, the protocol was sharpened step by step, and today we can presume that the world has managed to solve the ozone problem at the last second. Yet it will still take many decades before the ozone hole has closed. This demonstrates once more how inertly the climate reacts, and that forward-looking action is demanded.

8.2 Greenhouse Gas Emissions

The emission and the concentration of the long lasting greenhouse gases have risen dramatically in the last decades, especially since World War II. In spite of our knowledge about the anthropogenic influence exerted on the climate, even today there is no sign that the global emissions of greenhouse gases are sinking. As illustrated in Figure 27, despite the Convention on Climate Change of Rio de Janeiro and the Kyoto Protocol, global carbon dioxide emissions have actually risen dramatically since 1990.

While (energy-related) emissions were still about 22 billion tons in 1990, by 2007 they already amounted to around 31 billion tons, an increase of 40%. Particularly striking in this is the fact

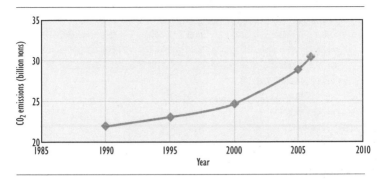

Figure 27   The evolution of global energy-related CO_2 emissions
1990–2007 (billion tons)

that emissions are increasing exponentially, with the rate of increase actually accelerating. It is plausible to assume that this trend will not be reversed within the coming years, so that we will continue to face steep increases in CO_2 concentrations. A glance at the emissions broken down by countries (Figure 28) shows that above all the United States and China, with a combined share of around 40%, emit far more than all other countries.

The tremendously dynamic economic development in threshold countries like China and India will make it especially difficult to achieve a significant reduction in the emissions of greenhouse gases in the coming decades. The enormous reserves of coal in China are a particular cause for concern. But even within the European Union (EU) many countries have increased their CO_2 emissions. In 2004, Spain and Portugal, for instance, recorded an increase of over 40% compared to 1990. The strongest reductions, of over 50% since 1990, were achieved by the three Baltic states Estonia, Latvia and Lithuania. Considerable reductions of about 30% were achieved in Poland, Hungary, Slovakia and the

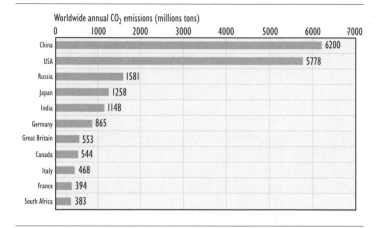

Figure 28 The energy-related CO₂ emissions (million tons) per country
          in the year 2007

Czech Republic as well. With a reduction of 20% between 1990 and 2007, Germany ranks in the middle of the table. The EU as a whole probably will not manage to reduce CO_2 emissions within the next years, as demanded by the Kyoto Protocol.

8.3 Options for Action

These numbers document that it will be tremendously difficult to reduce the emissions of greenhouse gases in the short term. Therefore it is reasonable to assume that emissions will continue to rise, and, that, if a global reduction in the emission of greenhouse gases can be realized at all, it will take several decades. So is the situation already hopeless? To answer this question, the inertia of the climate must be taken into consideration. Moreover, it is also important to consider the inertia of the economy. Economically sensible strategies must attempt to avoid both severe climate changes and major upheaval for the global economy.

At this juncture it makes sense to take a look at the time horizon of the next thousand years. We will use two typical BAU ('business as usual') scenarios proposed by the IPCC, designated with the letters C and E (Figure 29), and use them as drivers for a simplified climate model that allows long time periods to be simulated within a short time. Here CO_2 emissions to the atmosphere rise under the assumption that all available fossil resources are burned. In the first simulation (curve C), 4000 Gt C are burned in total (note that the emissions are given here in units of carbon (C), not in units of carbon dioxide (CO_2)). This corresponds more or less to what has been estimated as the reserves of conventional fossil fuels (oil, natural gas, coal). In the second simulation the estimated emissions are still higher, at 15 000 Gt C (curve E), because the assumption in this scenario was that 'exotic' reserves, like heavy oil and shale oil or tar sand, will also be burned.

The concentration of CO_2 rises in the next centuries to values between 1200 ppm in scenario C and 4000 ppm in scenario E, i.e. to a multiple of the preindustrial value of 280 ppm. The two simulations render the climate up to the year 3000. Such calculations are very helpful for the understanding of the climate system, as they demonstrate fundamental aspects of the climate's reaction. Yet it should also be mentioned that using such extreme scenarios can mean departing from the models' scope of application. Thus caution is due in any detailed interpretation of the results.

CO_2 concentrations remain relatively high during the entire simulated millennium. This is the case even for scenario C, in which the emission of CO_2 sinks to zero in approximately 2300, after which humans release no more carbon dioxide into the atmosphere. Nevertheless in the year 3000 the CO_2 concentration

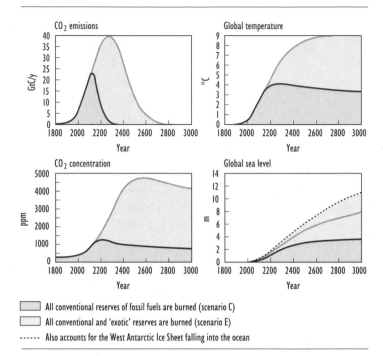

☐ All conventional reserves of fossil fuels are burned (scenario C)

☐ All conventional and 'exotic' reserves are burned (scenario E)

------ Also accounts for the West Antarctic Ice Sheet falling into the ocean

Figure 29 Development of CO₂ emissions (Gt C/year) over time, CO₂
          concentrations (ppm), global temperature (°C) and global sea
          level (m) under the assumption of two BAU scenarios

          In scenario C all conventional reserves of fossil fuels are burned, in scenario E
          the 'exotic' ones are burned as well. The dotted line in the figure for sea level
          also accounts for the West Antarctic Ice Sheet falling into the ocean. Most of the
          climate changes do not begin until the second half of this millennium. All
          changes are represented as deviations from their preindustrial values

is still far above 500 ppm. The situation is even more extreme in
scenario E, with a CO₂ concentration of around 4000 ppm at the
end of this millennium. Obviously, the Earth system is very slow
to rid the atmosphere of the carbon dioxide introduced by man.

The picture is similar for the climate's reaction seen in terms of global temperatures and the global sea level. In both scenarios the global mean temperature remains at a very high level even after the reduction of CO_2 emissions, and cools off only very slowly. Even more extreme is the situation for the sea level, which rises due to a combination of thermal expansion, the melting of the mountain glaciers, and the gradual melting of the Greenland Ice Sheet. The dotted line at the top also accounts for the West Antarctic Ice Sheet breaking off into the ocean. Not even at the end of this millennium has a new equilibrium been achieved. Sea level rises especially intensively in the second half of this millennium, with values that can exceed several meters. Even in the year 3000 sea level continues to rise. Thus it is altogether possible that we will leave a tremendous rise in sea level behind for a great many successive generations.

A climate change – like higher temperatures, a rise in sea level, or changes in the frequency of droughts and flooding – has direct effects on us humans and on the ecosystems. We will either adapt to the climate change or undertake steps to change our patterns of action. This, in turn, will result in changes in the emission of greenhouse gases to the atmosphere, and thus in its influence on our climate. Therefore only an integrated approach using models of the Earth system makes sense, in which we calculate not only the climate, but also the social and economic aspects of global climate change. In order to prevent grave climate changes, the global emission of greenhouse gases would have to decline dramatically from today's levels within the next one to two hundred years. Yet there is no need to scramble. The climate is inert and will react only to a long-term strategy. This means that it will no longer be possible for us to influence the development of the climate over the next decades in any fundamental way. Today we must forge the path for the development of the climate thereafter.

This can be illustrated through two more simulations, shown in Figure 30. In this model the costs of climate change are compared with the corresponding 'abatement costs.' If, for instance, new technologies are introduced in order to reduce the emission of greenhouse gases, this requires financial expenditures. Inversely, the damages caused by climate change, for example through flooding or droughts, also cost money. Thus it makes sense to calculate 'optimum' emission paths that take both kinds of costs into consideration and represent a compromise between climate and economy. A change in the global economy takes time, too, as otherwise costs would skyrocket. Power plants and machines have typical life cycles of several decades, and they cannot be substituted overnight. This would not make economic sense and would result in major economic disruptions, which are not in anyone's interest. Hence it is more reasonable to integrate a certain economic inertia into the calculations. The advantage of this report is that it facilitates the development of an optimum strategy that does not place too great a burden on the global economy in the short term, but nevertheless preserves the climate as much as possible.

While it is possible to reduce emissions by 50% immediately without taking economic inertia into consideration, when this factor is considered, emissions actually continue to rise for a few decades before falling significantly. The interesting result of comparing the two simulations is that the calculated climate changes do not differ from each other very much. Apparently what counts is only that emissions are reduced significantly in the long term. A crash course would not be terribly effective and is thus not required. The decisive goal must be to significantly reduce greenhouse gas emissions in the long term, i.e., over a period of around one hundred years, and ultimately bring their level down to zero. Although the two strategies are very different in the

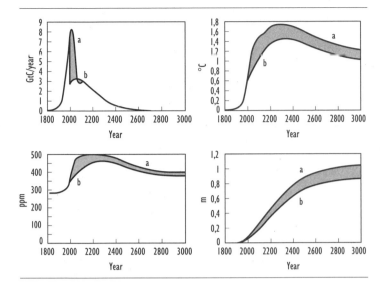

Figure 30 Optimum courses of action calculated with (curve a) and
without (curve b) consideration of economic inertia for this
millennium

Depicted are the model's optimized values for CO_2 emissions (GtC/year), the
corresponding concentrations (ppm), global mean temperature (°C) and global
mean sea level (m). All variables are shown as deviations from their
pre-industrial values

next two to three decades, this has hardly any effect on climate
change. In both simulations it is possible to keep the temperature
change under 2°C, while the rise in sea level stabilizes at about 1
m. These changes may not be minor, but for the most part they
can no longer be prevented. Thus we should work together with
industry to develop the long-term strategy necessary to protect
both the climate and the economy from major disruptions.

Because of the long residence time of CO_2 in the atmosphere, the climatic response is governed by the cumulative CO_2

emissions rather than by the detailed path. Short-term measures over a period of several years thus play practically no role for the climate. This also explains why the Kyoto Protocol alone cannot achieve the climate protection required. Considering the climate's long reaction times, a five-percent decrease in the emission of greenhouse gases by the year 2012 is inconsequential in practice. Only if emissions continue to decline consistently after 2012 will the content of greenhouse gases in the atmosphere stabilize and gradually decrease. Therefore it does not make any sense for states to play number games in jostling for seemingly advantageous positions. Whether emissions are reduced by 5% or only 4% is meaningless for the long-term development of the climate. Hence it is high time to cease and desist with the haggling about percentages at climate conferences. Instead we should keep our sights on the long-term nature of the climate problem and develop long-term strategies for climate protection that can be supported by all countries. In this a particularly important role falls to the introduction of renewable energies.

8.4 How Do We Deal with the Climate Problem?

Since the inertia of the climate means that it basically reacts only to a long-term strategy, it is not too late to launch an effective climate protection process that avoids serious climate changes. Important in this is the development of renewable energies, above all of solar energy. These new technologies can be developed gradually within the coming decades and then applied without any major detriment to the economy. The next climate conferences should serve as opportunities to agree on the modalities of how to encourage the breakthrough of renewable energies worldwide. In the short term, efforts should concentrate on saving energy and using it more efficiently, in order to make a contribution to the goals of the Kyoto Protocol now. There is still tremendous unused potential for energy savings, especially in the industrialized nations.

These demands are directed to every country, as the climate problem is one faced by the whole world. Greenhouse gases do not respect international boundaries and always have a global effect, no matter where they are released. In so far it is important that all states participate in climate protection, especially the two greatest emitters of greenhouse gases, the United States and China. These two states still adamantly refuse to join the Kyoto process. The story of the circumstances of the ozone hole is in many ways exemplary for the way we humans have dealt with environmental problems. Although we often recognize them at an early stage, we fail to approach these problems courageously and solve them. Apparently we do not wake up until something serious happens. It seems that reason alone is not enough; first a certain pain threshold must be reached. Discussions between the different interest groups drag on endlessly, paralyzing political decision makers. Environmental groups warn, while industry fails to recognize the problems.

Yet we also see this hesitation in many other fields of politics, especially here in Germany. Be it health or social security policy, we address our problems much too late. The changed age structure of our society, for instance, is not something we just discovered yesterday. This is similar to the way the climate problem is dealt with on the global political stage. It is also typical that everyone is for protecting the environment when asked, but only the fewest act accordingly. The multitude of SUVs on our roads speaks for itself. We are also not willing to pay higher prices for environmentally friendly products. The endless discussions concerning the price of gasoline speak volumes. The word sustainability is heard over and again, but neither citizens nor their representatives actually behave in a sustainable manner. A fundamental change in consciousness is thus required. Sustainable behavior must become a matter of fact for all of us.

Petroleum, in particular, as the most important fossil fuel, embodies in exemplary manner the irresponsible way we deal with the environment. During its extraction the oceans and soil are contaminated. It is transported over distances of thousands of kilometers in tankers that do not meet even the most rudimentary security standards and are operated by poorly trained crews working for rock-bottom wages. Thus it is no surprise that every year sees at least one serious accident involving oil tankers. Over and again, entire coastal regions are devastated by oil spills, not to mention the damages under water, which we do not even see. The images of marine birds daubed with oil are broadcast around the world. We are shocked and distraught, but two weeks later the topic has disappeared from the media and our minds. Then everything continues as before until the next accident. And in the end we burn the oil, thus releasing into the atmosphere billions of tons of the greenhouse gas carbon dioxide each year, and polluting it further with other contaminants like aerosols. No longer can we treat our Earth as a garbage dump. The climate problem is only one of many environmental problems to be addressed; we are poisoning not only our air, but also the ground and the oceans. The latter are becoming ever more acidic, as they absorb increasing amounts of the carbon dioxide we pump into the atmosphere. The pollution of the environment is such a gigantic challenge for world politics that it raises the question as to whether it is at all possible to master our environmental problems. It would be important and helpful if a personality were to enter the world stage who made an issue of the environment and set in motion a revolution in environmental policy.

Such a personality must recognize that the time has come to act, in spite of all uncertainties that still exist and always will. Many politicians are fond of claiming that more research is necessary before any truly accurate predictions of global climate change can be submitted. In the United States, especially, the political strategy has been to point out the uncertainties and shirk any decisions. However, it is erroneous to expect that the case of absolute certainty will ever arise. The very nature of predictions on global climate change makes them uncertain. Yet the consequence must not be to put off dealing with the climate problem. We know enough about the climate problem today and the precautionary principle alone demands that we react. We simply owe it to the many generations that will follow us. In so doing we must keep in mind that there may be surprises we cannot yet conceive of today. For this the ozone hole serves as a cautionary example. Others include Germany's extremely hot summer of 2003, and the extraordinarily strong hurricane season over the Atlantic in 2005, which were not expected in such intensity, even accounting for global climate change.

We know that the concentration of carbon dioxide has not been as high as it is today at any time in the last 800 000 years. Both experimental and theoretical physics showed long ago that carbon dioxide and other trace gases absorb thermal radiation from the Earth with the result that the Earth's surface and the lower layers of air heat up. The temperature trend in recent decades can no longer be explained by natural processes alone; to an increasing extent, man is determining the global climate. After all, it is undisputed that, should the concentrations of greenhouse gases continue to rise so rapidly, the Earth will heat up to an unprecedented degree. These are hard facts, agreed upon by all scientists. Uncertainties exist, of course. Yet these are of secondary importance and certainly do not justify our failure to implement measures to protect the climate. Consequent action is advisable, from absolutely everyone involved.

Nevertheless, let us address the uncertainties once more. There are three main factors that make climate change predictions uncertain. First is the question of how mankind will behave in the future. If we significantly curb the global emission of greenhouse gases, we can expect that climate change will end up rather minor. But if we continue to increase emissions, we must plan on quite a strong climate change. A second reason is the chaotic nature of the climate system. Because of this, climate predictions always are of a probabilistic nature. We can only assign a certain probability to the occurrence of certain changes; there is no absolute certainty. Yet this source of uncertainty is minor in comparison to the one resulting from our ignorance about the future emissions of greenhouse gases. And finally, the climate models themselves are not perfect. Errors are incorporated, yet these errors are not glaring enough to render the models worthless. The mere fact that the Earth will heat up to a degree it has never done since humans appeared on the planet is important enough. That is why it does not make much sense to argue about whether the temperature on Earth will warm up by 3°C or 4°C. This is a purely academic discussion, and one that certainly must be conducted in the fields of research. But for politics this question is of no consequence, because in the view of most experts, the Earth should be prevented from warming up by more than 2°C relative to the preindustrial era.

So we must accept the uncertainty and learn to live with it. This is sometimes difficult to communicate. We humans place great value on certainty; we like to have clear alternatives between which we can choose. Yet each of us knows that we have to deal with uncertainties in every life situation, we just do not always realize it. Whenever we drive a car there is a small probability of experiencing a traffic accident, yet we suppress this risk. The same is true every time you go out the door: a shingle could fall on your head. Implicitly, our entire life is characterized by the fact that we assess probabilities and behave accordingly. The same is true for the climate problem. The probability that we have already changed the climate is well over 90%. In everyday life we would regard such a high probability as certainty. Thus it is all the more astonishing that we insist on complete certainty where the climate is concerned. It must be said once more, loud and clear: as far as it is humanly possible to tell, we are in the process of initiating a climate change unlike any ever seen before, which will last longer than a millennium and pose tremendous problems for many successive generations.

But we must not exaggerate. Not everything in the way of storms that happens on this planet is caused by us humans. The media are fond of seizing on extreme weather events in order to draw attention to global climate change. This is especially true for tropical cyclones like hurricanes. These storms are certainly very destructive and bring much suffering upon the people affected. But the damages caused by hurricanes are also increased by the fact that ever more people live in areas at risk. Moreover, we must not forget that the world always experienced hurricanes. Even a critical review of the observations of the last 150 years has not revealed any statistically significant long-term trend, neither upward nor downward, although 2005 was the year with the most hurricanes, and also brought the strongest hurricane ever observed since measurements began. The problem is the strong natural fluctuations of the ocean surface temperature in the tropical Atlantic, which, in turn, influence the hurricanes. This does not mean that hurricanes cannot change as a consequence of global warming, or may have done so already. An intensification or increased frequency of hurricanes would even be extremely plausible, because global warming has already caused the ocean temperature in the tropical Atlantic to rise. In this connection the clustering of very strong hurricanes in recent years is grounds for concern and should be understood as a warning sign. Yet from the scientific perspective it is not proven that we humans have significantly changed the statistics of hurricanes.

Not all weather processes necessarily have to change simultaneously as a consequence of global warming. Inversely, though, the fact that some components have not changed yet or will not change at all does not argue against the existence of global climate change. One of the important factors that determines when changes can be proved to be caused by us humans is the 'signal-to-noise' ratio. We regard a certain change to be unusual only if it departs from the range of natural fluctuation. And even then it must be shown that man is actually the cause. What is more, we must not forget that we are still at the very first stage of global climate change. Many phenomena will not change until warming becomes more intense, and some may react with more inertia than others. This is why it makes no sense to portray the situation in black and white. Some things have already changed demonstrably through us humans, and others will change in the future, but there will also be aspects of weather that do not change at all. A differentiated consideration of the climate problem is thus advisable. Simple arguments that pick out only one special process or aspect of the problem are not suitable for answering the question of how humans influence the climate.

But how can the inert Earth system be managed? We are familiar with many examples of inert systems. For example, an express train has a very long braking distance. This means that braking has to begin several kilometers before the train station. The same is true for ships: piloting gigantic tankers requires great foresight. In order to simplify the control of the climate,

Earth system models can be developed. In contrast to pure climate models, Earth system models take into consideration not only the physical and biogeochemical aspects, but also the social components, above all the global economy. Unfortunately such models are still in the fledgling stages of development, but they already provide some interesting results that could help guide important global political decisions. In this connection it is important to communicate that reactions can unfold gradually and thus the necessary restructuring of the global economy does not have to be completed within just a few years.

While the United Nations Framework Convention on Climate Change of Rio de Janeiro of the year 1992 is, in principle, a declaration of intent, the Kyoto Protocol of the year 1997 is more concrete. Its implementation would mean the first time in recent years that the rapid rise of greenhouse gases in the atmosphere was significantly curbed.

Yet the Kyoto Protocol still excludes developing countries. Since the start of industrialization, it has been chiefly the industrial nations that have filled the atmosphere with greenhouse gases. If anything like morals or justice exists in world politics, then we, the people in the developed countries, must begin to reduce greenhouse gas emissions. We can not simply pass this responsibility on to the developing countries, which have contributed very little to the climate problem so far. If we are really serious about protecting the climate, then we in the industrial nations have to make a start. The developing countries also have an interest in climate protection, for they are the ones that will probably have stronger consequences of global climate change to deal with. Yet they will still want to continue developing, and they have every right to do so. The only way out of this trap is for us in the industrial nations to develop efficient, or, even better, entirely new technologies to acquire energy, and then to provide this to the developing countries. This is especially important as regards the development of technologies for exploiting solar energy. Many developing countries have an abundance of the natural resource sun. We could thus kill two birds with one stone: on the one hand, we could protect the global climate through the increased use of solar energy; on the other, we would offer many developing countries a solid prospect for economic development, as solar energy would provide them with an important natural resource. Along this route it might be possible to achieve a situation that closes the great gap in living standards between the industrial nations and the developing countries, easing the North-South conflict. Then all countries would profit from the development of technologies to exploit renewable energies. In so far we would then be on the way to a more just world.

Leaving the climate up to the Kyoto Protocol alone would have a negligible influence. The point now is to tighten up its stipulations further at the annually scheduled follow-up conferences. The ozone problem can serve as an example here. The original Montreal Protocol was also quite soft, but at subsequent conferences it was tightened so far that today the prospects are promising. Thanks to its policy measures, we now can proceed from the assumption that the ozone hole will be closed again within the next 50 years. The problem of global warming is much more complex than the ozone problem, however, because in practice we have hardly any alternatives to fossil fuels at this time. Solving the climate problem thus first necessitates massive investments in research and technology, in order to make it possible for the global economy to shift toward renewable energies in the long term.

The great inertia of the climate also presents an opportunity. It would not make much difference if it took a few years before the United States or China committed to climate protection and then began emitting fewer trace gases into the atmosphere. At the climate conferences the long-term nature of the climate problem must be acknowledged, and considerations turn to the question of how the global economy can convert to a carbon-free economy over the course of the next decades. This is in the interest of us all, as fossil energies are ultimately limited and our oil reserves may last for only another 50 years. We should contemplate how the breakthrough of renewable energies can be encouraged across national boundaries. The short-term economic interests that currently constitute an obstacle for a number of states to ratify the Kyoto Protocol play a secondary role on the longer time scales of decades or even centuries. This is why it also makes sense to accommodate Kyoto deniers like the United States, for in the long term we have the same interests, whether we like it or not, namely breaking our dependence on fossil energies. Hardened fronts do not help anyone; the community of states can only solve the problem by working together.

Now and again several countries make the impression that they are banking on being less affected by climate change. Yet the example of the Indonesian Tambora volcano, which erupted in 1815 and was the most violent eruption in modern history, shows how global climate problems really are. Even distant areas like North America and Europe were affected by extreme cold in the subsequent years, which led to famines and political unrest. There will be no winners in global climate change; we will all lose. This is what the climate models show all too clearly. An increase in extreme weather events, for example, will affect all countries, and the rise in sea level will affect all coasts.

Germany could take on an exemplary function in climate protection. We have the financial means and the technological know-how to do so. Germany is well on its way. We have already reduced our CO_2 emissions by about 23% since 1990. About half of this is due to reunification, when the antiquated technology in the new federal states was replaced by modern plant. Yet the other half of these savings was 'real,' and came about through such measures as improved energy exploitation and the deployment of renewable energies. Today the state of Schleswig-Holstein, for instance, generates about 30% of its energy through wind.

Many of us still remember the oil crisis of 1973, when the supplies of petroleum suddenly dried up. Since the oil crisis an interesting trend can be observed, even in the United States. While up to this point the emissions of CO_2 ran hand in hand with the development of the gross domestic product, the most common measure of a country's economic strength, these values have diverged ever since: since the oil crisis CO_2 emissions have risen considerably more slowly than economic power. A similar trend has been observed in Japan. Germany has even reduced its emissions in recent years. This allows only one conclusion: the myth of economic stagnation when energy consumption is reduced cannot be maintained. Yet neither may we make the mistake of believing that the climate problem can be mastered by merely saving more energy or improving energy efficiency. The global population will grow, many countries will become more industrialized, and thus mankind's demand for energy will grow considerably in the next decades. Therefore in the long term there is only one path we humans can take. Against the backdrop of the climate problem and shrinking fossil resources, we must generate energy in some way other than by burning fossil fuels.

If we could make only a fraction of solar energy utilizable, we would no longer need petroleum, natural gas, or coal. Germany, as a country without natural resources, must set store in innovation. Fossil energy is growing more expensive by the day, and we will experience how fossil energy continues to rise in cost in the coming years. Anyone who takes the development of alternative technologies seriously now will have an economic advantage in the future. For this reason policy should encourage investments directed toward the future and support more vigorous efforts to bring renewable energies to market. Multinational oil concerns recognized this long ago and are already performing research on solar energy. BP, for instance, advertises a stronger utilization of renewable energies with the slogan 'beyond petroleum'. Corresponding signals are also coming from German energy suppliers. In the industrial sector a process of rethinking has apparently begun. Therefore politics and economy should now put their heads together to find steps to lead us out of the energy crisis.

In Germany we are well on our way, but we could do much better. We still hold the largest share in greenhouse emissions within the European Union. However, already a good 10% of German electricity consumption is provided by renewable energies. According to its own statistics, the sector has now grown to revenues of 16 billion euros and 160 000 employees. From 2004 to 2005 the share of electricity from wind, water, sun and biomass climbed from 9.4% to 10.2%. In 2005 a total of 62 billion kilowatt hours of green electricity was fed into the German power grid. In 1999 the Renewable Energies Law was enacted, according to which energy providers have to accept green electricity at fixed rates. Based on total energy consumption in Germany – including heat as well as electricity – renewable energies amounted to 4.6% in 2005 (2004: 4.0%). As such the goal of 4.2% of primary energy demand set for 2010 has already been exceeded. The main source of green energy is still wind power, which accounts for 26.5 billion kilowatt hours, followed by hydropower (21.5 billion) and biomass (ten billion). One billion kilowatt hours comes from solar power. The goal of the government is to increase the share of renewable energies on the electricity market to at least 20% by 2020.

We citizens enjoy great power. We should not forget that we owe it to our children and grandchildren and their children to leave them an intact environment. Therefore the environmental revolution must also come from below. Climate protection begins with each and every one of us. Only if we grasp this can we Germans serve as a positive role model. Everyone should consider what he or she can do to protect the climate. Saving energy is an important step in this direction. Be it by driving less often or by improving heat insulation in our homes, by avoiding waste or by turning off stereos and televisions rather than leaving them on – everyone can make a contribution. One problem is that the costs of destroying the environment are borne not by the culprits, but by the world at large. Anyone who would like to drive a SUV can do so without being held accountable for the subsequent costs. As long as such products are available for less than their true price, without incorporating the cost of damage to the environment, they continue to enjoy a competitive advantage over alternative, ecologically friendly products. We citizens can change this by buying products that put less of a strain on the environment. If we do so consistently, industry will adjust to our behavior at lightning speed. Unfortunately, the German automobile industry is not fulfilling its own obligation to significantly reduce the consumption of its vehicles. Here pressure by consumers would be helpful to remind the producers of their promise. But political leaders must also understand that voluntary obligations by industry are not always successful. If these are not fulfilled, statutory regulations should be considered.

Our consciousness must also change. We should admire not the SUV drivers who unduly pollute the environment, but rather the people who protect it by driving an economy car. This should be true for companies, too, of course. The tropical rainforest can serve as a cautionary example. One of the greatest scandals of environmental policy is that the destruction of the tropical forests still continues. By now around two thirds of the rainforest in Indonesia have been cut down. Every two seconds an area the size of a soccer field is destroyed. The slash and burn there not only produces hefty CO_2 emissions, it also robs many animals of their habitat; every day species disappear from the planet, never to return. We are thus destroying important genotypes and assaulting biological diversity. The example of the rainforest shows that economic interests often do not take the interests of the environment into consideration. However, the consumer could contribute to halting the deforestation of the tropical rainforest by refusing to buy the goods it is cut down to produce.

So, each and every one of us can contribute something to the environment and protection of the climate. This is important, because only then can we, the Federal Republic of Germany, take on a leading role. We should not think that the individual is powerless. On the contrary. However, policy must reward those who behave in an environmentally friendly manner. Since we live in a capitalized world, this reward must also pay off in euros and cents. There are positive approaches. For instance, schools in Hamburg are instituting the 'fifty/fifty' program to encourage pupils and teachers alike to save energy. The effects are noticeable not only in their reduction of CO_2 emissions, but also in savings on heating costs and other overheads. The school is allowed to keep half of this financial gain; the other half goes back to the city. In this way all participants profit: the schools, the city of Hamburg, and, of course, the environment. Schools, in particular, gain a certain flexibility through the 'fifty/fifty' program, because they can use the money to buy things that previously seemed beyond their means. This produces tremendous motivation for both pupils and teachers to use energy as sparingly as possible.

Germany is a country of exporters. Innovation is our livelihood. The current economic crisis demonstrates clearly that we have arrived at a cul-de-sac. Our economy continues to be relatively strong on the international scale; for instance, we still build good cars that are in high demand abroad. But this is no longer enough. Other countries are catching up and are now building cars that are just as good, which several countries can produce at considerably lower cost. And several countries have already overtaken us. It is incomprehensible why the hybrid automobile comes not from Germany, but Japan. Hybrid technology, i.e., the combination of an internal combustion engine and an electric engine, is a promising energy-saving propulsion technology, which converts the energy normally wasted during coasting into electric power. The demand for hybrid cars is already quite high. This example demonstrates how countries that develop efficient technologies have a competitive advantage in today's globalized world. The army of the unemployed here in Germany speaks for itself: we simply cannot produce at low wages. We will never be able to compete with labor costs in countries like Korea or China. Hence there is no point in even joining this battle. Apparently the German automobile industry, especially, has yet to recognize what kinds of cars will be demanded in the future. Unfortunately, its focus is still too strong on size and too weak on intelligent technology.

One of the most important challenges of coming decades is the energy question. This is becoming clearer and clearer, first because of the skyrocketing cost of fossil energies, and second because of political upheaval. The latter is manifest at the moment in the Iran crisis. Those who are not at the forefront in the development of renewables will lose their economic competitiveness completely. We Germans must thus have a direct interest in pioneering this field. Of course we will continue to build automobiles in the future, but their propulsion must be revolutionized. There are first positive steps in this direction, like the fuel cell, for instance. Yet to avoid the emission of greenhouse gases into the atmosphere and protect the climate, the hydrogen needed for this technology must be produced using a renewable method. This is not enough. Of course it will be easy to exploit solar energy in the desert. Many companies are sitting on plans prepared to do just this. But we have to think further. We in Germany, too, receive energy from the sun the whole year round. The challenge is therefore not how to make use of the sun in the desert, but rather how we can extract energy from our modest sunshine. We have to bank on innovation, develop new technologies and export them.

A great role in this is also played by education, both in schools and at the university. Innovation depends on the citizens of a country. Unfortunately, in Germany there is not yet any research landscape that occupies itself intensively with renewables. Therefore investment in the development of this sector must be increased.

Innovation is the only way to guarantee welfare, and the only way to solve the climate problem. We should not build on the many 'engineering solutions' propagated from various quarters. One proposal, for instance, is to channel carbon dioxide into the deep sea. But one should always address the root of the problem. For global warming this means that the emission of greenhouse gases into the atmosphere must be reduced. The Earth is an extremely complex system. Whenever we interfere with this system, we are confronted with new surprises. If we were to channel carbon dioxide into the deep sea, then the life forms down there might be endangered. The organisms that live deep down in the ocean may not be as highly developed as here on the Earth's surface, but we are beholden to this life nonetheless.

Moreover, there presumably will be other effects we cannot even conceive of today. Others demand that certain substances be introduced into the atmosphere in order to increase cloud cover, which is then supposed to cool the Earth further. And others again would like to shoot gigantic mirrors into space to divert solar radiation. One can only warn against undertaking such experiments. We should not believe that we have understood the Earth system so well that we can deal with it like any everyday object. We should eliminate the disturbance we humans have caused as quickly as possible. Then the Earth system, according to everything we know today, will recover gradually, i.e. over the course of several decades.

The climate problem can be solved. There is still time to act. Although we can no longer prevent some degree of global warming on the Earth, the extreme changes are still preventable. This demands something that is very rare in democratic societies. We all have to pull together: politics, industry and citizens – in Germany, but also globally. The climate problem is also a symptom of other unresolved global policy problems. And this, in turn, means that we ultimately have to solve all global political problems together. Sustainability must be understood in this overarching context. Only if we monitor our actions with regard to their consequences on all components of the Earth system will it be possible to live in harmony with nature.

Glossary

absorption – Matter can 'swallow' electromagnetic radiation, which generally leads to warming. Particularly important is the absorption of long wave radiation by certain atmospheric gases, which is the foundation for the greenhouse effect.

aerosols – Suspended sediment in the atmosphere. Aerosols have important influence on the radiation balance of the atmosphere. Aerosols introduced into the atmosphere by humans have a cooling effect and thus dampen the anthropogenic greenhouse effect.

albedo – The capacity a material (molecules, clouds, land surface, ocean surface, etc.) possesses to reflect incident solar radiation. Light areas have a high albedo, dark surfaces have a very low albedo.

atmosphere – The gaseous mantle surrounding the Earth. The atmosphere of a planet has a decisive influence on its climate. The surface of Venus is very hot, for instance, because most of Venus' atmosphere consists of carbon dioxide and thus exhibits a very strong greenhouse effect.

biogeochemical feedback – The physical climate system consisting of atmosphere, ocean and cryosphere (ice sphere) interacts with the other components of the Earth system (like vegetation, carbon cycle, etc.). The feedback mechanisms these components exert on the physical climate system are called biogeochemical feedback. So, for instance, the CO_2 content of the atmosphere was relatively low during the ice ages, and thus the greenhouse effect reduced, which amplified the cooling effect.

biosphere – A component in the Earth system, which exerts an important influence on the climate. The biosphere can influence the climate directly (e.g. through the albedo of vegetation) or indirectly via feedback mechanisms with the materials cycles (e.g. as a source or sink for carbon dioxide).

Brownian motion – From theoretical physics it is known that large particles can be induced to slow movements through coincidental collisions with small particles. This principle of Brownian motion can be translated to the climate system, which consists of components with very different time scales. Thus, for instance, high-frequency (fast) random weather fluctuations can be integrated by the more inert components (e.g. the ocean) and result in low-frequency (long-period) fluctuations of the ocean surface temperature.

CFCs – Chlorofluorocarbons are artificial products that enter the atmosphere only through human agency and can destroy the ozone layer. CFCs also contribute to the anthropogenic greenhouse effect and thus to global warming of the atmosphere.

cirri – Cirri (singular: cirrus) are high clouds consisting of ice particles, which are very thin and thus can be permeated by solar radiation. Yet cirri also absorb strongly in the long-wave range of the spectrum. An increase in cirrus cloud coverage would thus result in a warming of the Earth's surface and the lower layers of the atmosphere.

climate models – Using the laws of physics it is possible to create an image of the Earth that can be used to conduct experiments. The totality of all physical equations and parameterizations that describe the development of the climate system is called a climate model. Because of the complexity of the corresponding mathematical equations, these are solved by approximation using the methods of numerical mathematics and supercomputers.

climate subsystems – The climate system consists of different components called climate subsystems. The most important of these are atmosphere, ocean, cryosphere and biosphere. The various climate subsystems have very different time scales and are closely interconnected with each other.

condensation – Water can occur in various states. The phase transition from water vapor to liquid water is called condensation. The heat released during the process of condensation is called 'heat of condensation.' Condensation plays an important role in cloud formation.

Coriolis force – One of the effects of the Earth's rotation on large-scale motions. Winds in highs and lows flow parallel to isobars because the force of the pressure gradient is largely balanced by the Coriolis force. This phenomenon is known as geostrophic balance.

cryosphere – The climate subsystem that contains all ice on the Earth. The components of the cryosphere are sea ice, ice sheets, shelf ice, mountain glaciers, and surfaces covered with snow.

El Niño – The strongest short-term natural climate fluctuation. El Niño is manifested as a large-scale warming of the equatorial Pacific, and returns every four years on average, evoking climate anomalies worldwide. The El Niño phenomenon is part of a cycle called the El Niño/Southern Oscillation (ENSO). The cold phase of the cycle is called La Niña. ENSO is predictable and represents a breakthrough in seasonal forecasting.

emission – The term emission is used in two different respects. First, the output of greenhouse gases by humans is called anthropogenic emission. Second, depending on their temperature, materials emit electromagnetic radiation, a process that is also known as emission.

fingerprint method – A complex statistical procedure to identify the influence of humans on the climate. Analogous to criminalistics, this method takes advantage of the fact that every influence on the climate, natural and anthropogenic, exhibits a special time-space structure. Fingerprinting methods of this kind are already in use today to prove humans' influence on the climate.

fossil fuels – The fossil fuels are petroleum, natural gas and coal. These formed millions of years ago. Burning fossil fuels for the purpose of generating energy is the most important anthropogenic source of carbon dioxide.

Gleissberg cycle – The incident sunlight hitting the Earth is subject to fluctuations. On the decadal time scale, the most important fluctuation is the Gleissberg cycle, with a period of approximately 80 years and an estimated amplitude of approximately 0.2–0.3% of the solar constant.

greenhouse effect – Certain atmospheric trace gases absorb and emit electromagnetic radiation in the thermal range of the spectrum and thus lead to an additional warming of the Earth's surface and the lower layers of air. The natural greenhouse effect has a magnitude of about 33°C. Humans increase the concentration of certain trace gases relevant for the climate, like carbon dioxide, thus reinforcing the greenhouse effect, the result of which is global warming.

greenhouse gases – The trace gases involved in the greenhouse effect. The most important greenhouse gas for the natural greenhouse effect is water vapor, with a share of around

60%. For the 'anthropogenic' greenhouse effect caused by us humans, carbon dioxide plays the most important role, responsible for about 60%.

ice age cycles – Over the course of millennia the history of the Earth has experienced strong climate swings. The most prominent fluctuations are the ice ages. Different cycles can be identified, with periods of ca. 100 000, 41 000 and 19/23 000 years. These are induced by gradual changes in the parameters of the Earth's orbit, which exhibit precisely these periods. Positive feedback mechanisms in the Earth system reinforce each other, like, for example, the equiphase changes in the concentrations of greenhouse gases.

ice calving – Pieces constantly break off of ice sheets, falling directly into the ocean. This 'calving' process may be accelerated by global warming, thereby contributing to sea level rise. Ice calving also influences the salt content of the ocean and is thus important for the general circulation of the ocean as well.

ice clouds – Ice clouds are located at high altitudes and are also called cirri. The condensation trails generated by aircraft are ice clouds of this kind.

interglacial – In the history of the Earth severe climate changes have taken place. The colder phases are known as ice ages, the warmer ones as interglacials. We are currently in an interglacial period, the Holocene. The last major interglacial was the Eemian interglacial about 125 000 years ago.

IPCC – The Intergovernmental Panel on Climate Change was founded in 1988 by the United Nations Environment Programme (UNEP) and the World Meteorological Organization (WMO), to document the state of scientific knowledge on climate research, and to consult on global policies. Many hundreds of the world's leading climate scientists contribute to the IPCC reports. The IPCC reports (the most recent of which is from the year 2007) are considered to be the most reliable specialized reports on the subject of global climate change.

Kyoto Protocol – The Kyoto Protocol was passed in 1997. It prescribes that the industrial nations must reduce their emission of greenhouse gases by an average of 5.2% compared to 1990 over the period from 2008 – 2012. The Kyoto Protocol took effect in February 2005 upon its ratification by Russia.

La Niña – The cold phase of the El Niño/Southern Oscillation (ENSO) phenomenon. La Niña events are expressed in an abnormal cooling of the equatorial Pacific and, like El Niño events, exert climate effects worldwide.

Lorenz model – A simple, non-linear conceptual model with three components, on the basis of which several characteristic features of chaotic systems, such as weather, can be studied.

meridional temperature gradient – The temperature contrast between the tropics and the poles. The meridional temperature gradient is a crucial factor determining the structure of general atmospheric circulation.

Milankovitch theory – The theory explaining ice age cycles. The Earth's path around the sun and the characteristics of the tilt of the Earth's rotation axis are subject to long-period oscillations that influence the Earth's climate.

Montreal Protocol – An international pact for the protection of the ozone layer, which was passed in 1987. Together with the stricter stipulations passed since then, the Montreal Protocol regulates the emission of CFCs. There is hope that with the Montreal Protocol the long-term recovery of the stratospheric ozone layer can be achieved.

movement of the Earth's crust – The Earth's crust is not a solid shell; it is broken up into huge, thick plates that drift atop the soft, underlying mantle. Furthermore, the Earth's crust can sink under the pressure of large ice sheets and rise again after this ice melts. Certain areas of the Scandinavian Peninsula, having been released from the ice masses of the last ice age, are currently rising by up to one meter every hundred years.

ocean acidification – Some of the carbon dioxide (CO_2) emitted by us humans is taken in by the oceans, which leads to the acidification (a drop in the pH value) of the ocean waters. This effect is already measurable. All told, between 1800 and 1995, the oceans absorbed about 48 per cent of the cumulative CO_2 emissions from fossil fuels (including cement production), or 27–34 per cent of the total anthropogenic CO_2 emissions (including those from land-use changes. Since the start of industrialization the pH value has already fallen by about 0.11 units.

ozone – Ozone is triatomic oxygen. Most ozone occurs in the stratosphere (above an altitude of around 15 km). The ozone layer there absorbs the UV radiation dangerous for living beings, so that it arrives on the Earth's surface in a much weaker intensity. Humans also produce ozone, especially during typical smog weather situations in the summer. This ground-level ozone is not to be confused with stratospheric ozone.

ozone hole – The extensive thinning of the ozone layer over the Antarctic, observed every year at the beginning of spring on the Southern Hemisphere (spring in the Southern Hemisphere is from September through November).
The cause of the ozone hole can be traced back to man.

Anthropogenic emission of CFCs is the main reason for the destruction of the ozone layer.

parameterization – The approximative (numerical) solution of the mathematical equations describing the evolution of the climate system is performed on a calculation grid. Physical processes with scales smaller than the mesh width of the grid, like cloud formation, for example, cannot be resolved explicitly and therefore have to be represented using the information available at the grid points (parameterized). Typical grid widths in global climate models run to just a few hundred kilometers. The parameterization of sub-scale processes represents one of the most important sources of uncertainty in climate models.

polar stratosphere clouds (PSCs) – Ice clouds that can form in the stratosphere at temperatures of −78°C and lower in the pole regions. They play a central role in the development of the ozone hole over the Antarctic.

polar vortex – A very stable wintertime circulation regime over the poles, which prevents the exchange of air masses. The polar vortex over the Antarctic is particularly stable, such that it provides ideal conditions for the formation of polar stratosphere clouds.

range of the solar spectrum – Every body emits electromagnetic radiation corresponding to its temperature (see 'emission'). The wavelength of the maximum of radiation is dependent on the temperature. The higher the temperature of a body, the shorter the wavelength of the emitted radiation. The very hot sun thus emits primarily short-wave radiation in the visible range. The main range of frequencies emitted by the sun is called the solar spectrum range.

rheology – The study of the deformation and flow of matter. One of rheology's tasks is to determine how ice behaves under the influence of forces like wind stress.

Schwalbe cycle – The 11-year sunspot cycle that changes the period of solar radiation that hits the Earth. The amplitude of the 11-year sunspot cycle amounts to about 0.1% of the total solar irradiation.

shelf ice – Extensions of the ice sheets on land can protrude well into the ocean. The part of such an ice sheet that floats on the water is called shelf ice.

solar constant – The solar radiation incident at the upper edge of the Earth's atmosphere per unit of surface. The solar constant has a value of 1367 W/m². The solar constant is subject to periodic oscillations (Schwalbe and Gleissberg cycles).

stratosphere – In terms of its vertical temperature profile, the Earth's atmosphere can be divided into stories. The lowest story, up to an altitude of 10 – 15 km, is known as the troposphere. Located over this is the stratosphere, which extends to a height of about 50 km. The stratosphere contains the ozone layer so important for life on Earth.

thermal range of the spectrum – In comparison to the sun, temperatures on the Earth are relatively cold. The electromagnetic radiation emitted by the Earth thus is located primarily in the non-visible, infrared range of the spectrum, which is called the thermal range of the spectrum.

thermohaline circulation – An oceanic conveyor belt that transports large amounts of heat. In the high northern latitudes (Greenland Sea, Labrador Sea), cold and salty (and thus dense) water masses sink, streaming toward the Equator at great depths. Warm water flows along the surface toward the north. This revolving movement is called the thermohaline circulation. It is of paramount importance for the climate of northern Europe. The Gulf Stream is part of the thermohaline circulation.

topography – The relief of the ocean floor or an ice sheet. The topography of the ocean is an important boundary condition for oceanic currents; the topography of ice sheets an important one for the atmosphere located above them.

trace gases – The main components of the Earth's atmosphere are nitrogen (78%) and oxygen (21%). Because of their low concentrations, the remaining gases are called 'trace gases.' A number of these trace gases, like water vapor and carbon dioxide, exert a decisive influence on the terrestrial climate (greenhouse effect).

trade winds – A system of winds directed from the subtropics to the Equator. The closer the Equator is, the stronger the westward component of the trade winds. The trade winds near the Equator play an important role in the interaction between ocean and atmosphere and for the emergence of El Niño and La Niña events.

troposphere – The lowest story of the atmosphere, which extends up to altitudes of around 10 – 15 km, depending on the latitude. The important weather phenomena take place in the troposphere.

water vapor feedback – An initial change in atmospheric temperature results in a change in the concentration of water vapor. Because water vapor is a greenhouse gas, this change has implications for the temperature. Water vapor feedback is generally positive, i.e., it has a reinforcing effect. As the consequence of global warming triggered by mankind, water vapor content in the atmosphere will also increase, since the capacity of air to absorb water vapor increases with temperature. More water vapor

means a stronger greenhouse effect; thus water vapor
feedback reinforces the effect of anthropogenic emissions of
greenhouse gases.

weather – The fluctuations of the state of the atmosphere on
time scales of minutes up to several days.

Bibliography

Bundesministerium für Bildung und Forschung: Herausforderung Klimawandel. Available from the BMBF, Postfach 30 02 35, 53182 Bonn, 2003.

Bjerknes, J.: Atmospheric teleconnections from the equatorial Pacific. In: Mon. Wea. Rev. 97, pp. 163–172; 1969.

Dijkstra, H. A.: Nonlinear Physical Oceanography. A dynamical systems approach to the large-scale ocean circulation and El Niño. Dordrecht, London, Norwell, New York 2000.

Hasselmann, K.: Stochastic climate models. Part I: Theory. In: Tellus 28, pp. 473–485; 1976.

Hurrell, J. W.: Decadal trends in the North Atlantic Oscillation: Regional temperatures and precipitation. In: Science 269, pp. 676–679; 1995.

IPCC 2001a: Climate Change 2001. The Scientific Basis. Working Group I. Available from the IPCC, c/o WMO, C. P. 2300, 1211 Geneva, Switzerland, 2001.

IPCC 2001b: Climate Change 2001: Synthesis Report. Available from the IPCC, c/o WMO, C. P. 2300, 1211 Geneva, Switzerland, 2001.

IPCC 2007a: Climate Change 2007. The Physical Science Basis. Working Group I. Available from the IPCC, c/o WMO, C. P. 2300, 1211 Geneva, Switzerland, 2001.

IPCC 2007b: The AR4 Synthesis Report. Available from the
    IPCC, c/o WMO, C. P. 2300, 1211 Geneva, Switzerland,
    2001.

Latif, M. and T. P. Barnett: Causes of decadal climate
    variability over the North Pacific and North America. In:
    Science 266, pp. 634–637; 1994.

Lorenz, E. N.: Deterministic nonperiodic flow. In: J. Atmos. Sci.
    20, pp. 130–141; 1963.

Lozan, J., Graßl, H. and Hupfer, P. (eds.): Das Klima des 21.
    Jahrhunderts. Warnsignal Klima. Available from J. Lozan,
    Büro Wissenschaftliche Auswertungen, Schulterblatt 86,
    20357 Hamburg, 1998.

Palmer, T. N.: A nonlinear dynamical perspective on climate
    change. In: Weather 48, pp. 314–326; 1993.

Palmer, T. N.: Extended-range atmospheric prediction and the
    Lorenz model. In: Bull. Am. Met. Soc. 74, pp. 49–65; 1993.

Philander, S. G.: Is the temperature rising? The uncertain
    science of global warming. Princeton, New Jersey 1998.

Schmincke, H.-U.: Volcanism. Berlin, 2004.

Schönwiese, C.-D.: Klimatologie. Stuttgart 2003.

Stommel, H.: Thermohaline convection with two stable regimes
    of flow. In: Tellus 13, pp. 224–230; 1961.

Sverdrup, Sc. U.: Wind-driven currents in a baroclinic ocean
    with application to the equatorial currents of the eastern
    Pacific. In: Proc. Nat. Acad. Sci. 33, pp. 318–326; 1947.

Timmermann, A., Latif, M., Voss, R. and Grötzner, A.:
    Northern Hemispheric Interdecadal Variability: A coupled
    air-sea mode. In: J. Climate 11, pp. 1906–1931; 1998.

Umweltbundesamt. Klimaänderung. Festhalten an der
    vorgefassten

Meinung. Wie stichhaltig sind die Argumente der Skeptiker?
    Berlin, 2004.

Visbeck, M. et al.: Atlantic Climate Variability Experiment
    (ACVE), Prospectus 1998. *www.ldeo.columbia.*
    edu/~visbeck/acve
WBGU (Wissenschaftlicher Beirat der Bundesregierung Globale
Umweltveränderungen). Sondergutachten: Die Zukunft der
    Meere – zu warm, zu hoch, zu sauer, pp. 67–76; 2006.

Climate modeling

Promet, 2002, Jahrgang 28: 'Numerische Klimamodelle – Was
    können sie, wo müssen sie verbessert werden? Teil I: Das
    Klimasystem der Erde.'
Promet, 2003, Jahrgang 29, Heft 1–4: 'Numerische
    Klimamodelle – Was können sie, wo müssen sie
    verbessert werden. Teil II: Modellierung natürlicher
    Klimaschwankungen.'

Popular science books

Crutzen, P. J. and Müller, M.: Das Ende des blauen Planeten?
    Der Klimakollaps. Gefahren und Auswege. München 1997.
Grassl, H. (ed.): Wetterwende. Vision: Globaler Klimaschutz.
    Frankfurt a. M./New York 1999.
Grassl, H. and R. Klingholz: Wir Klimamacher. Auswege aus
    dem globalen Treibhaus. Frankfurt am Main 1990.
Latif, M.: Hitzerekorde und Jahrhundertflut. Herausforderung
    Klimawandel. München 2003.
Stevens, W. K.: The change in the weather. People, Weather,
    and the science of climate change. New York 1999.

Web pages

www.dkrz.de
For images and animations on the subject of weather and climate

www.hamburger-bildungsserver.de
General information on the subject of climate

www.ipcc.ch
Reports of the Intergovernmental Panel on Climate Change

www.umweltbundesamt.de
Information on climate protection. Discussion of several skepticists' arguments

Sources of Figures

All graphics: Peter Palm, Berlin. In color section: Figure 1: NASA; Figure 2: ©picture-alliance/dpa; Figure 3a: © Sammlung Gesellschaft für ökologische Forschung; Figure 3b: © Gesellschaft für ökologische Forschung; Figure 4: NASA.